IET CONTROL ENGINEERING SERIES 95

Modelling Control Systems Using IEC 61499

Other volumes in this series:

Volume 8	**A history of control engineering, 1800–1930** S. Bennett
Volume 18	**Applied control theory, 2nd edition** J.R. Leigh
Volume 20	**Design of modern control systems** D.J. Bell, P.A. Cook and N. Munro Editors)
Volume 28	**Robots and automated manufacture** J. Billingsley (Editor)
Volume 33	**Temperature measurement and control** J.R. Leigh
Volume 34	**Singular perturbation methodology in control systems** D.S. Naidu
Volume 35	**Implementation of self-tuning controllers** K. Warwick (Editor)
Volume 37	**Industrial digital control systems, 2nd edition** K. Warwick and D. Rees Editors)
Volume 39	**Continuous time controller design** R. Balasubramanian
Volume 40	**Deterministic control of uncertain systems** A.S.I. Zinober (Editor)
Volume 41	**Computer control of real-time processes** S. Bennett and G.S. Virk (Editors)
Volume 42	**Digital signal processing: principles, devices and applications** N.B. Jones and J.D.McK. Watson (Editors)
Volume 44	**Knowledge-based systems for industrial control** J. McGhee, M.J. Grimble and A. Mowforth (Editors)
Volume 47	**A history of control engineering, 1930–1956** S. Bennett
Volume 49	**Polynomial methods in optimal control and filtering** K.J. Hunt (Editor)
Volume 50	**Programming industrial control systems using IEC 1131-3** R.W. Lewis
Volume 51	**Advanced robotics and intelligent machines** J.O. Gray and D.G. Caldwell (Editors)
Volume 52	**Adaptive prediction and predictive control** P.P. Kanjilal
Volume 53	**Neural network applications in control** G.W. Irwin, K. Warwick and K.J. Hunt (Editors)
Volume 54	**Control engineering solutions: a practical approach** P. Albertos, R. Strietzel and N. Mort (Editors)
Volume 55	**Genetic algorithms in engineering systems** A.M.S. Zalzala and P.J. Fleming (Editors)
Volume 56	**Symbolic methods in control system analysis and design** N. Munro (Editor)
Volume 57	**Flight control systems** R.W. Pratt (Editor)
Volume 58	**Power-plant control and instrumentation: the control of boilers and HRSG systems** D. Lindsley
Volume 59	**Modelling control systems using IEC 61499** R. Lewis
Volume 60	**People in control: human factors in control room design** J. Noyes and M. Bransby (Editors)
Volume 61	**Nonlinear predictive control: theory and practice** B. Kouvaritakis and M. Cannon (Editors)
Volume 62	**Active sound and vibration control** M.O. Tokhi and S.M. Veres
Volume 63	**Stepping motors, 4th edition** P.P. Acarnley
Volume 64	**Control theory, 2nd edition** J.R. Leigh
Volume 65	**Modelling and parameter estimation of dynamic systems** J.R. Raol, G. Girija and J. Singh
Volume 66	**Variable structure systems: from principles to implementation** A. Sabanovic, L. Fridman and S. Spurgeon (Editors)
Volume 67	**Motion vision: design of compact motion sensing solution for autonomous systems** J. Kolodko and L. Vlacic
Volume 68	**Flexible robot manipulators: modelling, simulation and control** M.O. Tokhi and A.K.M. Azad (Editors)
Volume 69	**Advances in unmanned marine vehicles** G. Roberts and R. Sutton (Editors)
Volume 70	**Intelligent control systems using computational intelligence techniques** A. Ruano (Editor)
Volume 71	**Advances in cognitive systems** S. Nefti and J. Gray (Editors)
Volume 72	**Control theory: a guided tour, 3rd edition** J. R. Leigh
Volume 73	**Adaptive sampling with mobile WSN** K. Sreenath, M.F. Mysorewala, D.O. Popa and F.L. Lewis
Volume 74	**Eigenstructure control algorithms: applications to aircraft/rotorcraft handling qualities design** S. Srinathkumar
Volume 75	**Advanced control for constrained processes and systems** F. Garelli, R.J. Mantz and H. De Battista
Volume 76	**Developments in control theory towards glocal control** L. Qiu, J. Chen, T. Iwasaki and H. Fujioka (Editors)
Volume 77	**Further advances in unmanned marine vehicles** G.N. Roberts and R. Sutton (Editors)
Volume 78	**Frequency-domain control design for high-performance systems** J. O'Brien
Volume 81	**Optimal adaptive control and differential games by reinforcement learning principles** D. Vrabie, K. Vamvoudakis and F. Lewis
Volume 88	**Distributed control and filtering for industrial systems** M. Mahmoud
Volume 89	**Control-based operating system design** A. Leva et al
Volume 90	**Application of dimensional analysis in systems modelling and control design** P. Balaguer
Volume 91	**An introduction to fractional control** D. Valério and J. Costa
Volume 92	**Handbook of vehicle suspension control systems** H. Liu, H. Gao & P. Li

Modelling Control Systems Using IEC 61499

2nd Edition

Alois Zoitl and Robert Lewis

The Institution of Engineering and Technology

Published by The Institution of Engineering and Technology, London, United Kingdom

The Institution of Engineering and Technology is registered as a Charity in England & Wales (no. 211014) and Scotland (no. SC038698).

First published 2001
Second Edition 2014

The Institution of Engineering and Technology
Michael Faraday House
Six Hills Way, Stevenage
Herts, SG1 2AY, United Kingdom

www.theiet.org

British Library Cataloguing in Publication Data
A catalogue record for this product is available from the British Library

ISBN 978-1-84919-760-1 (hardback)
ISBN 978-1-84919-761-8 (PDF)

Typeset in India by MPS Limited
Printed in the UK by TJ International Ltd, Cornwall

Contents

Foreword ix
Foreword xi
Preface xiii
Abbreviations and conventions xvii

1 Introduction 1
1.1 IEC 61499 function block standard 5
1.2 Development of function block concept beyond IEC 61131-3 8
1.2.1 Global variables 10
1.2.2 Communications function blocks 11
1.3 Development of IEC 61499 12
1.4 Why use function blocks 14
1.5 System design views 16
1.5.1 Logical view 17
1.5.2 Process view 17
1.5.3 Development view 18
1.5.4 Physical view 18
1.5.5 Scenarios 18
1.6 The future beyond IEC 61499 19

2 IEC 61499 models and concepts 21
2.1 Function block model 22
2.1.1 General characteristics 22
2.1.2 Execution model for function blocks 24
2.1.3 Function block types 27
2.2 Application model 28
2.3 System model 29
2.3.1 Overall system structure 29
2.3.2 Device model 30
2.3.3 Resource model 30
2.4 Distribution model 32
2.4.1 Mapping of applications 32
2.4.2 Platform specific configurations 34
2.5 Management model 34
2.5.1 Management applications 36
2.5.2 Operational state model of managed entities 37

2.6	Exchange format for IEC 61499 entities	39
	2.6.1 Textual syntax for IEC 61499 entities	39
	2.6.2 XML-exchange format	41
2.7	Summary	43
3	**Defining function block types**	**45**
3.1	Types and instances	45
3.2	Different form of function blocks	45
3.3	External interface declaration	46
	3.3.1 Static interface declaration	47
	3.3.2 Defining the dynamic interface behaviour	48
	3.3.3 2 out of 3 voter example	50
3.4	Defining basic function blocks	53
	3.4.1 Internal behaviour	53
	3.4.2 Execution control chart features	55
	3.4.3 Execution of basic function blocks	56
	3.4.4 Behaviour of instances of basic function blocks	59
	3.4.5 Basic function block examples	59
3.5	Definition composite function blocks	62
	3.5.1 Rules for composite function block type specification	62
	3.5.2 Execution of composite function blocks	64
	3.5.3 Composite function block example	64
3.6	Summary	68
4	**Structuring applications**	**69**
4.1	Application structuring with subapplications	69
	4.1.1 Rules for subapplication type specification	71
	4.1.2 Rules for subapplication execution	72
	4.1.3 Subapplication example	72
	4.1.4 Subapplication distributed example	74
	4.1.5 Guidelines for using subapplications	77
4.2	Structured interfaces using adapters	78
	4.2.1 Adapter concept	79
	4.2.2 Defining adapter types	81
	4.2.3 Modelling the behaviour of adapter types	82
	4.2.4 Using adapter types in function blocks	83
	4.2.5 Adapter interface usage example	85
	4.2.6 Guidelines for using adapters	88
4.3	Summary	88
5	**Service interface function blocks**	**91**
5.1	Overview	91
5.2	Interface specifications	93
	5.2.1 Standard inputs and outputs for service interface function blocks	93

5.2.2 Service sequence diagrams for service interface function blocks 94
5.3 Type definitions 98
5.3.1 General service interface function block types 98
5.3.2 Example service interface function block types 100
5.3.3 Textual syntax – service interface function block example 101
5.4 Communication service interface function blocks 103
5.4.1 Function blocks for unidirectional transactions 103
5.4.2 Function blocks for bidirectional transactions 104
5.5 Management function blocks 106
5.5.1 Examples 108
5.5.2 Managed function blocks 108
5.6 Summary 109

6 Event function blocks 111
6.1 Overview 111
6.2 Standard event function block types 113
6.2.1 Event Splitter 114
6.2.2 Event Merger 114
6.2.3 Two Event Rendezvous 114
6.2.4 Permissive Event Propagator 116
6.2.5 Event Selector 116
6.2.6 Event Switch 117
6.2.7 Delayed Event Propagator 117
6.2.8 Restart Event Generator 118
6.2.9 Cyclic Event Generator 118
6.2.10 Event Train Generator 119
6.2.11 Event Train Table-driven Generator 120
6.2.12 Separate Event Table-driven Generator 121
6.2.13 Event-driven Bistable 122
6.2.14 D (data latch) Bistable 123
6.2.15 Boolean Rising Edge Detector 123
6.2.16 Boolean Falling Edge Detector 124
6.2.17 Event-driven Up Counter 125
6.3 Using event function blocks for event flow control 125
6.3.1 Basic event flow control 125
6.3.2 Conditional event propagation 127
6.3.3 Modelling control tasks 128
6.4 Summary 129

7 Industrial application examples 131
7.1 Overview 131
7.2 Temperature control example 132
7.3 Conveyor test station example 135
7.3.1 Distributed system model 141

7.4 Fieldbus applications 143
7.4.1 Analogue input function block example 146
7.4.2 Further generalisation of the analogue input function block 151
7.5 Concluding remarks 153
7.6 Summary 154

8 Epilogue **155**
8.1 Current status of IEC 61499 155
8.2 Engineering Support Task 157
8.3 Compliance to IEC 61499 159
8.3.1 IEC 61499 compliance profiles 159
8.3.2 Device classes 160
8.4 Large-scale industrial applications 161
8.5 Summary 162

Appendix A Common elements **165**

Appendix B IEC 61499 Compliance Profile for Feasibility Demonstrations **175**

Appendix C Frequently asked questions (IEC 61499 FAQ) **181**

Appendix D PID function block example **191**

Appendix E Exchange formats **207**

Appendix F Bibliography **219**

Index **221**

Foreword

In early 1990, Technical Committee 65 of the International Electrotechnical Commission (IEC TC65) initiated a project to develop a common architectural model for the application of software modules called ‘function blocks’ in distributed industrial-process measurement and control systems (IPMCS). This model was to encompass IPMCS utilising ‘Fieldbus’ systems as specified in IEC 61158 as well as IPMCS utilising the programming languages defined in Part 3 of the IEC 61131 standard for programmable controllers. This project resulted in the current IEC 61499 standard for function blocks.

Due to the relative immaturity of the IEC 61158 project at the time of the proposal, experts and a project leader were not available for the 61499 project until approximately two years after its inception, at which time the first edition of IEC 61131-3 was also completed and available for reference. Because of the close relationship between IEC 61131-3 and the projected IEC 61499, many of the experts participating in the development of the latter came from the previous 61131-3 project, including Bob Lewis and myself.

Through a long process of systematic application of software engineering and open systems principles, with intensive international review and revision, the TC65/WG6 experts reached consensus on the basic concepts and detailed technical approach to the resolution of a number of fundamental issues. This resulted in the publication in 2002-4 of IEC/PASs (Publicly Available Specifications) 61499-1, -2 and -4, for a two-year period of trial implementations, as well as IEC/TR 61499-3 containing tutorial information. This initial consensus was reflected and thoroughly explained in the first edition of the present book.

In the years to follow, the accumulation of experience in the adoption and implementation of IEC 61499 resulted in the publication of 61499-1, -2 and -4 as IEC Standards in 2005, and in second edition in 2012. Alois Zoitl was a key member of the IEC working group during this period, as well as being one of the prime developers of the widely used 4DIAC open-source software tool and runtime platform for the development and implementation of IEC 61499 applications.

I would like to express my gratitude to Bob and Alois for bringing out this updated edition of Bob’s pioneering book on IEC 61499. It is my hope that this will provide the basic information needed to promote the ongoing industrial adoption of IEC 61499, as well as serving as a textbook for courses in advanced industrial process automation and control.

James H. Christensen
Cleveland Heights, Ohio USA
14 December 2013

Foreword

Is our birthday which we celebrate every year an event or a cyclic event?

Unlike in normal life, in industrial automation, we have a standard leading the way, namely IEC 61499.

The question of event-driven versus cyclic execution is a topic that is most discussed between the IEC 61131-3 and IEC 61499 representatives. However, in our experience, the customer is not interested in events or cycles, but in solutions to his problems in an increasingly networked, distributed and complex world. In addition it should be mentioned that the IEC 61499 and IEC 61131-3 standards are close relatives. They share the same roots but have a different scope and purpose. So someone who is familiar with IEC 61131-3 will also cope with IEC 61499 and appreciate it.

At the point of intersection, where an old technology is slowly being replaced by a new technology, it is most challenging. The cards are reshuffled and sometimes new opportunities arise. This is the time for pioneers, who look outside the box and develop ideas and upcoming solutions, resulting in a clear competitive edge. Anybody who, in addition, closely watches the industrial users, sharpens their view on real problems to develop their vision further. This is important if one is in the process of developing a product that relies on a new technology, which should be better than anything ever offered before.

Developing our vision and creating it as a marketable product was a long journey and at the beginning of this journey was the first edition of this book. It provides a solid foundation and is a good accompaniment for the practical implementation of one's ideas. The foundations are there, but the creativity in the implementation is, of course, the role of the system designer.

Already today we can claim and prove that IEC 61499 is the better solution for many tasks in automation, particularly for building automation, process automation, and machine control but also in the field of intelligent energy networks (e.g. smart grids based on IEC 61850). This is also reflected by the increasing interest in the market. This dynamic will bring a demand for people with expertise in the field of IEC 61499. With this growth in know-how, the spread and importance of the standard will increase at the same time. This book provides, therefore, the perfect start.

Finally, we would like to raise an appeal to current and future IEC 61499 vendors:

Keep compatibility high, in the long term it will be worthwhile for vendors and users alike!

Many thanks to all who supported, with their know-how, perseverance and passion, the development of the IEC 61499 standard and last, but not least, also to the authors of this book.

Horst Mayer (CTO nxtControl)
Arnold Kopitar (CEO nxtControl)
Leobersdorf, Austria
17 December 2013

Preface

New technologies and standards are emerging that are set to have a dramatic effect on the design and implementation of industrial control systems. These technologies will reduce the time to bring new systems onstream, and leverage the integration of business and industrial systems.

There is currently an explosion in the use of Object-Oriented (OO) software encapsulated as components. In business systems, the use of software components and technology together with service-oriented architectures is becoming more widespread. In industrial systems, programmable logic controllers (PLCs) and soft controllers based on Personal Computer (PC) architectures are also starting to adopt many of these techniques. The different worlds of factory automation and business systems are clearly starting to share the same software technology.

With so much scope for complexity new tools and techniques are clearly needed to design and model such systems. To date, new methodologies have emerged, such as the Unified Modelling Language (UML), which allow software engineers to deal with the complexity of OO-based systems.

Control and system engineers are also faced with increased software complexity as advances in industrial networking, such as Fieldbus and Industrial Ethernet technologies, allow intelligence to be distributed throughout the system from controllers, instruments, actuators and even out to the sensors themselves. As these systems become more complex, new tools and techniques are needed to model their behaviour.

Some of this complexity can be dealt with effectively by use of the IEC 61499 standard. It has been developed specifically as a methodology for modelling distributed industrial process measurement and control systems. This standard defines concepts and models so that software components in the form of function blocks can be interconnected to define the behaviour of a distributed industrial process measurement and control system.

Process and system engineers have used function blocks in various forms for a number of years as an effective way to represent and model software functionality in instrumentation and controllers. The PID algorithm is probably the best example of an early form of function block. New forms of smart devices and sensors now allow intelligence to be distributed widely throughout a control system. It is now becoming more difficult to define where the main intelligence of any control system really resides; intelligence is becoming truly distributed.

Tools based on the IEC 61499 standards are emerging that can model, validate and simulate the behaviour of complex function block networks. IEC 61499 defines a completely new development methodology, which requires a rethinking of control engineers. It has been in development for a number of years by international experts working in the field of control system software. It is set to become an important methodology for the process and system engineers working with complex distributed systems.

The main objective of this book is to widen the understanding of the most important concepts in the IEC 61499 standard and to show how these concepts can be applied to industrial problems. IEC 61499 is a complex standard that defines many new concepts related to function blocks and the supporting architecture in a rigorous and thorough manner – as a consequence, the standard can be difficult to understand by people who read it for the first time. This book has therefore concentrated on the main concepts and has intentionally summarised or omitted some of the less significant details in the standard. For this reason, some topics such as the Textual Syntax for describing function block definitions have not been covered in great detail.

IEC 61499 provides, for the first time, a framework and architecture for describing the functionality in distributed control systems in terms of co-operating networks of function blocks. It is the authors' intention that the benefits of this standard should be understood by a wide audience; including not only people working in industrial control and also those with a general interest in methodologies for modelling distributed systems.

Major progress since the first edition

Since the publication of the first edition of the book, many things have happened in the world of IEC 61499. The most prominent are as follows:

- The standard IEC 61499 has gone through two major revisions – when it was initially published as international standard in 2005 and then again when it was revised in the second edition in 2012. Many concepts have been refined, errors corrected and ambiguities clarified.
- At the time when the first edition was written only one prototype software tool, which supported the evaluation of the models and concepts of IEC 61499, was available. Since then several tools and runtime platforms have been developed. Now free, commercial and open source implementations, allowing the development of distributed industrial process measurement and control systems, are available.
- In the last 14 years much research has been focused on topics of IEC 61499. Investigated topics have included the execution of IEC 61499 models, development methodologies, formal verification methods and communication aspects – just to name the most prominent. This research has completely changed and improved how we now understand and use IEC 61499.
- We have been teaching IEC 61499 to people from industry and master level students for more than ten years. The questions and discussion with the students has resulted in a much deeper understanding of the standard and how it should be taught. From this we have learnt that the concepts of IEC 61499 should be introduced in a different order than was originally presented in the first edition.

Organisation, structure and changes in the second edition

With this second edition, we have incorporated our experience from the last 14 years and restructured the contents as follows:

Firstly, Chapter 1 gives an overview on the domain of industrial automation systems and their requirements. It aligns IEC 61499 with IEC 61131-3 covering software concepts such as object orientation and component orientation.

The main models and concepts of IEC 61499 are introduced in Chapter 2. This is the chapter that has undergone the most changes in order to highlight the interrelation of the different models. The concepts of the function block model and its use in applications has been put to the fore. This is followed by the system model describing the control hardware and the distribution model, which interlinks applications with the devices on which they will be executed.

Chapter 3 focuses on the main IEC 61499 function block types, i.e. basic and composite function block types. The descriptions of the types have been updated to the latest edition of the standard, common elements such as the interface definition have been consolidated and the examples have been expanded.

The new Chapter 4 describes means for structuring applications. The description of subapplications has been taken from the original Chapter 3, and adapters taken from the original Chapter 2. Their descriptions are significantly expanded and extended with new and reworked examples.

Chapter 5 describes service interface function blocks which provide means for interacting with services provided by devices or resources. Their description is aligned to Chapter 3 and a section on communication service interface function blocks has been added.

Chapter 6 gives an overview on event function blocks and how they can be used to control event flows in applications. The changes in this section are mainly corrections, but further examples and use cases of the event function blocks have also been added.

Chapter 7 shows how the models and concepts of IEC 61499 can be applied to implement control applications. The concluding Chapter 8 gives an overview on the current status of IEC 61499, including an outline of the requirements for software tools as given in IEC 61499 Part 2. Another addition in the second edition is an overview on compliance profiles as defined in IEC 61499 Part 4.

Acknowledgements first edition

The development of the IEC function block standard has taken many years involving numerous meetings, animated discussions, debate and argument. I must, therefore, thankfully acknowledge all contributions of the members of the IEC 65/WG6 working group whose efforts are distilled in the IEC 61499 standard. I would also particularly like to acknowledge the chairman, Dr. Jim Christensen who, through humour and gentle persuasion, ensured that the working group remained on track to create a concise and effective standard. I must also thank Jim and Rockwell Automation for the permission to use his FBDK tool, which I put to good use to prove the syntax and structure of the PID example given in Appendix D.

I would also like to thank Terry Blevins from Fisher-Rosemount Systems Inc. for all his information and help on the development of the Fieldbus standards. Some of his application examples have been particularly useful in demonstrating how function blocks can be used to model distributed systems.

I would also like to give special thanks to Dr. Bob Brennan at University of Calgary and Dr. John Wilkinson at Queen's University Belfast for all their help in reviewing the manuscript.

Bob Lewis, Worthing, UK, 2001

Acknowledgements second edition

First of all, I'm very grateful to Bob for the opportunity to write this second edition and his support in developing it. My thanks also go to Jim Christensen as he taught me the concepts of IEC 61499 back in 2001 and for all the discussions we had since then.

I would also like to thank the members and students of the Odo Struger Labor research group at the Automation and Control Institute, Vienna University of Technology, especially Reinhard Hametner, Ingo Hegny, Martin Melik-Merkumians, Carolyn Oates, Christoph Sünder and Monika Wenger, for our discussions on how to implement and apply IEC 61499. In this respect, I would like to thank also the research partners I had in this time, namely Franz Auinger, Gerhard Ebenhofer and Thomas Strasser.

I thank Jonathan Burdalo, and Benjamin Brandenbourger and Georg Neugschwandtner from fortiss GmbH for reading and commenting the manuscript of the second edition.

Last but not least my thanks go to my wonderful wife Andrea for her patience in the last months and her comments on many of the figures.

Alois Zoitl, Munich, Germany, 2014

Abbreviations and conventions

ASN. 1	Abstract Syntax Notation One
CPU	Central Processing Unit
DCS	Distributed Control System
FB	Function Block
FBD	IEC 1131-3 Function Block Diagram graphical language
FRD	Functional Requirements Diagram
GBs	GigaBits per second
HMI	Human–Machine Interface (see MMI)
HVAC	Heating, Ventilation and Air Conditioning
IEC	International Electrotechnical Commission
IPMCS	Industrial Process Measurement and Control Systems
I/O	Inputs and Outputs
IP	Internet Protocol
IS	International Standard
ISA	Instrument Society of America
ISO	International Standards Organisation
LD	IEC 1131-3 Ladder Diagram graphical language
MBs	MegaBits per second
MMI	Man–Machine Interface (now replaced by HMI)
MMS	Manufacturing Message Specification
OO	Object Oriented
OSI	Open Systems Interconnection
PAS	Publicly Available Specification
PC	Personal Computer
PID	Three term controller, Proportional, Integral, Derivative closed-loop control algorithm
P&ID	Piping and Instrumentation Diagram
PLC	Programmable Logic Controller
POU	IEC 1131-3 Program Organisation Unit

SCADA	Supervisory, Control and Data Acquisition system
SFC	IEC 1131-3 Sequential Function Chart graphical language
SI	Service Interface
SIFB	Service Interface Function Block
SOA	Service Oriented Architecture
ST	IEC 1131-3 Structured Text language
TCP	Transmission Control Protocol
UDP	User Datagram Protocol
UML	Unified Modelling Language
XML	eXtensible Markup Language

The following font and style is used to distinguish examples of textual programming language from ordinary text:

```
BEGIN
  A := A + 100.5; (* Comment *)
END
```

This text style is also used to reference textual programming language elements and also names used in the graphical representation in the text.

The suppression of superfluous detail in textual language examples is indicated using an ellipsis '...'.

Chapter 1
Introduction

In this introductory chapter we review the background and reasons behind the development of the IEC 61499 standard. Specifically we will:

- *review the design of current day control systems and consider the impact of new technology,*
- *look at the reasons for starting the development of the IEC 61499 standard,*
- *consider the reasons why function blocks are still an important concept to process and system engineers,*
- *show how function blocks have some of the characteristics of object- and component-oriented software,*
- *show how IEC 61499 models can be used in the control system development life cycle.*

As manufacturing companies fight to compete in today's unpredictable and ever changing global markets, there is an increased urgency to improve the agility of manufacturing systems. To produce competitive and innovative products, companies must be able to quickly design and create new forms of advanced automated production. Such levels of automation require the creation of large systems involving an amalgam of industrial control, manufacturing execution and business logistics systems. A key characteristic of all of these new systems will be a built-in ability to rapidly handle change, resulting in agile manufacturing systems [1]. A manufacturing plant will need to be able to quickly switch product types and bring new processes on stream to remain in business.

Currently there is growing interest in new technologies and architectures for creating the next generation of distributed systems for industrial automation. These will be systems where software is organised as sets of co-operating components rather than as an integration of large bespoke units of software [2].

Up to now, industrial control systems have been in one of two main camps, either based on traditional Distributed Control Systems (DCS) or based on Programmable Logic Controllers (PLCs). Current DCSs, as typically used in petrochemical plants and refineries, are structured around a few large central processors, which provide supervisory control and data acquisition, communicating via local networks with numerous controllers, instruments, sensors and actuators located out in the plant (Figure 1.1). A system may have both discrete instruments and out-stations with clusters of instruments with local controllers. In a DCS, the main supervisory

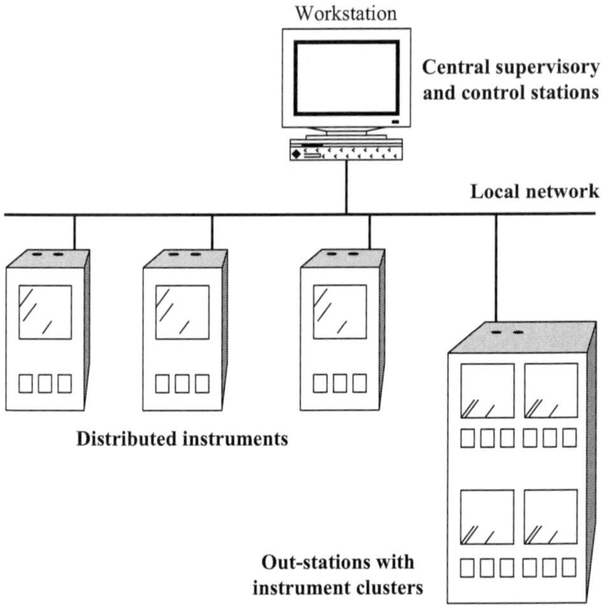

Figure 1.1 Distributed control system

control comes from one or more central processors. Instruments positioned out in the plant typically provide local closed loop control, such as for PID[1] control.

In contrast, for many machine control and production processes, particular on automotive production lines, systems have generally been designed using PLCs (Figure 1.2). In these systems, the Human–Machine Interface (HMI) is normally provided by a wide variety of different types of panels, lights and switches. Advanced HMI can also be provided by colour display panels with operator input via dedicated keypads or from touch sensitive screens. Especially with the introduction of smart phones and tablet computers this field is currently changing rapidly.

A large PLC system will generally have a number of PLCs communicating via one or more proprietary high-speed networks. PLCs will typically be connected to a large number of input and output (I/O) signals for handing sensors and actuators. In some cases, discrete instruments, for example, for temperature and pressure control, are also connected to PLCs.

With both design approaches, systems have tended to be developed by writing large monolithic software packages, which are generally difficult to reuse in new applications and are notably difficult to integrate with each other. Data and functionality of one application are not readily available to other applications, even if the applications are written in the same programming language and running in the same machine. Significant system development time is concerned with

[1] PID – Proportional, Integral and Derivative control algorithm commonly used to provide stable closed loop control (e.g. for temperature or pressure).

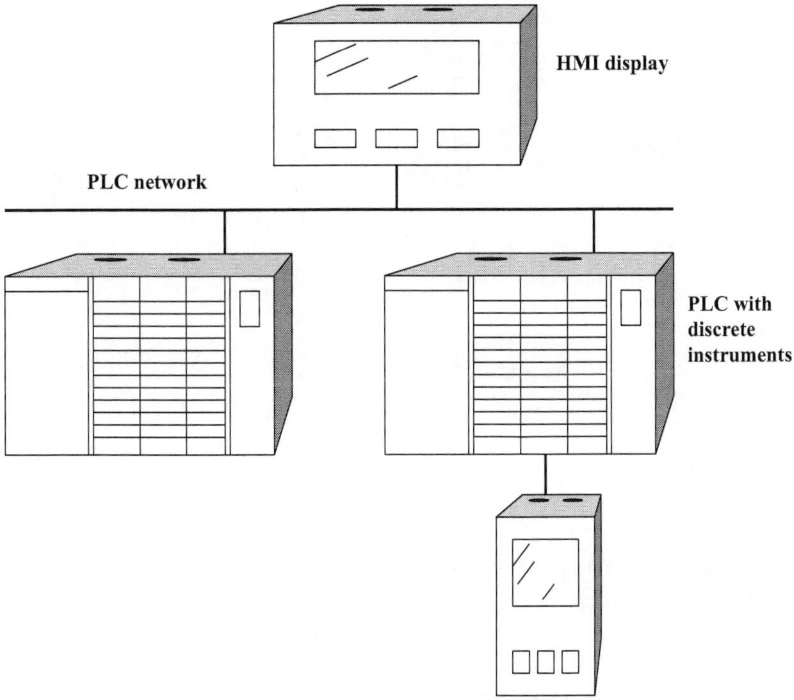

Figure 1.2 PLC control system

mapping signals between devices and providing drivers to allow different types of instruments and controllers to communicate.

Both types of system, DCS and PLC, tend to be difficult to modify and extend and do not provide the high degree of flexibility that is expected in systems for advanced and flexible automation.

With the emergence of standards in industrial communications such as Fieldbus, which will allow different types of instruments and control devices to interoperate, the differences between DCS- and PLC-based systems are starting to disappear. DCS instruments and PLCs are beginning to offer similar functionality. Industrial applications are also being implemented on PC hardware with concepts such as the SoftPLC, i.e. PLC logic running on a normal PC. As industrially hardened PC (IPC) platforms that offer high reliability become more common, we will see a trend to using more PC-based controllers. Until recently, classical PLCs could only be programmed using proprietary languages as offered by the PLC vendor. With users now requesting a more open approach to software, a new breed of soft-controller is emerging that can be programmed in a wide range of different programming languages. This new breed of soft-controllers is often referred to by the term Programmable Automation Controller (PAC).

We can now foresee the time when systems for controlling industrial, manufacturing and business processes start to merge. For example, a company can seamlessly link a business system running in a head office to manufacturing

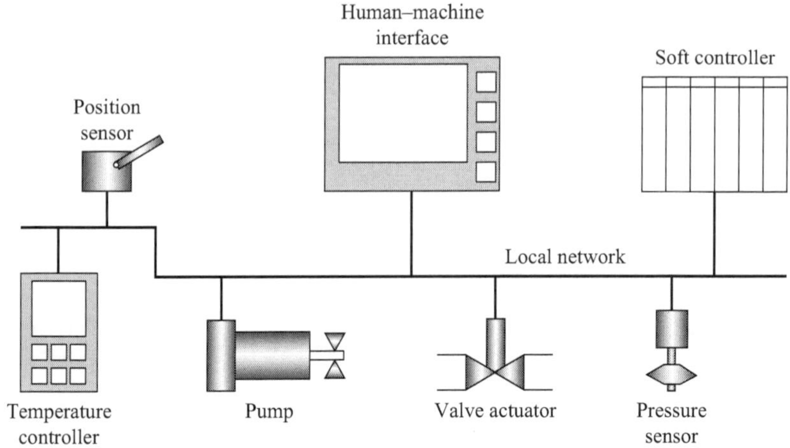

Figure 1.3 Advanced distributed functionality

processes and industrial control systems or even controllers running in any plant in any part of the world.

Figure 1.3 depicts part of a system having advanced distributed functionality. In such systems, each device connected to the industrial network can provide part of the control functionality. Smart devices, such as pumps, valves, or sensors will have built-in control functionality that can be linked by software with more intelligent devices such as HMI panels, temperature controllers and soft-controllers to form the total control system functionality. For example, a pressure sensor can be linked directly by software to a valve actuator and to a display bar graph on an HMI panel. A slider on an HMI panel can be directly software wired to the set-point of a PID controller controlling the speed of a pump.

To achieve these high levels of integration and yet enable the creation of flexible systems that can be re-engineered as industrial and business needs change, a completely new approach to software design will be required – a new technology based on the interaction of distributed objects [3]. There are several software technologies already well advanced that are set to have an influence in this area. The first have been middleware technologies like CORBA[2] [3] or DDS[3] [4] from the Object Management Group (OMG). A further decoupling was achieved by introducing Service-Oriented Architectures (SOA), where components offering services can be flexibly combined using these services [5]. On top of these technologies, several coordination technologies have been presented, e.g. Complex Event Systems [6] or the Enterprise Service Bus concept [7].

In the domain of industrial process measurement and control systems, the technologies from the OPC Foundation allow seamless data access regardless where they are located, be it in a remote industrial controller in a blast furnace or in the PC

[2] Common Object Request Broker Architecture.

[3] Data Distribution Service.

of the production manager's office [8]. Internet technologies using Ethernet, peer-to-peer communication or SOA are also being considered for manufacturing systems. With the new OPC Unified Architecture [9], there are proposals to bring this seamless interoperability also into small industrial devices (e.g. sensors and actuators).

The industrial community has long been aware that the ready interconnection of software components, such as in the form of function blocks, will have major advantages especially for end-users. These advantages will include improved software productivity through reuse of standard solutions, improve design flexibility by being able to plug-and-play software and devices from different vendors. So far, the new standards all enable 'technical integration' of distributed components, but the next major hurdle is 'semantic integration' (i.e. define a meaning behind the data). We may be able to link and exchange data between software in a remote industrial controller to a control algorithm running in a PC, but will the connection be meaningful?

1.1 IEC 61499 function block standard

The International Electrotechnical Commission (IEC) has developed a specific standard IEC 61499 [10] that defines how function blocks can be used in distributed industrial process, measurement and control systems. This work may help solve part of the semantic integration problem.

In industrial systems, function blocks are an established concept for defining robust, reusable software components. A function block can provide a software solution to a small problem, such as the control of a valve, or control a major unit of a plant, such as a complete production line. Function blocks allow industrial algorithms to be encapsulated in a form that can be readily understood and applied by people who are not software specialists. Each function block has a defined set of input data, which are read by the internal algorithm when it executes. The results from the algorithm are written to the function block's outputs. Complete applications can be built from networks of function blocks formed by interconnecting function block inputs and outputs.

The IEC 61499 standard, which builds on function block concepts defined in the PLC language standard IEC 61131-3 [11, 12], was developed in liaison with Fieldbus standardisation work. It is envisioned that the Application Layer part of the Fieldbus communications stack will provide the software interface to allow remote function blocks to interoperate over Fieldbus. However, IEC 61499 was developed as a generic standard that is also applicable in other industrial sectors – in fact, wherever there is a requirement for software components that behave as function blocks, for example in building management systems.

IEC 61499 defines a general model and methodology for describing function blocks in a format that is independent of implementation. The methodology can be used by system designers to construct distributed control systems. It allows a system to be defined in terms of logically connected function blocks that run on different processing resources.

Figure 1.4 depicts how the IEC 61499 methodology could be used as part of the system design life cycle. The design of a control system typically starts with the

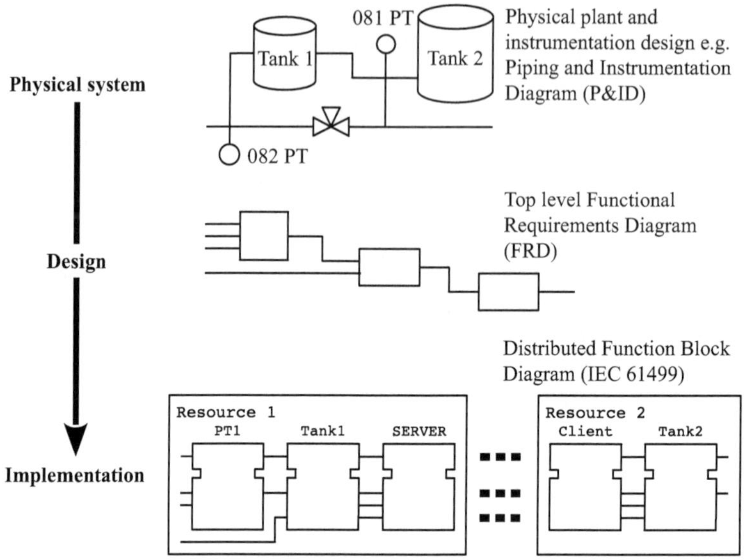

Figure 1.4 Applying IEC 61499

analysis of the physical plant diagrams and documentation on the control system requirements. This analysis leads to the phase of defining areas of functionality and their interaction with the plant. The final phase results in mapping functionality into physical resources such as PLCs, instruments and controllers.

The use of IEC 61499 can be best demonstrated by considering the following phases in the design of a distributed control:

- *Functional design phase*: During this phase, process engineers analyse the physical plant design, for example using Piping and Instrumentation Diagrams (P&ID), to create the top-level functional requirements. These can be represented as a series of blocks that outline the main software components and their primary interconnections. At this design phase, the physical distribution of the software blocks is not considered. In many cases diagrams that show the physical design of the plant or machinery, such as P&IDs, also show the location of active devices such as valves and pumps and instrumentation points, such as the location of pressure and temperature sensors.
- *Functional distribution phase*: In a distributed system, a further design phase is required to define the distribution of control functionality onto processing resources. IEC 61499 standard provides models and concepts for defining the distribution of functionality into interconnected function blocks. System engineers complete the detailed design by mapping the software requirements onto IEC 61499 function blocks. These may be distributed on various processing resources. In many cases, function blocks as provided in field devices will be exploited; for example intelligent devices such as smart valves may provide software packaged as a function block.

Each function block in turn will have its own particular software design life cycle. Some function blocks will need to be specifically designed for a system application, in other cases, existing function blocks within instrumentation and controllers can be used.

We will see later that the function block model defined by IEC 61499 provides just a subset of the views in a distributed system design. Other design views will be necessary to give all aspects of a system design. IEC 61499 is the first step in providing design methodologies for developing and modelling distributed applications.

As the trend to use component-based software continues, it is foreseen that industrial controllers and instruments will either provide function blocks as part of the device's firmware or provide function block libraries from which function blocks can be selected and downloaded. System design will become the process of software component selection, configuration and interconnection, just as much of electronic hardware design is now primarily concerned with the selection and interconnection of integrated circuits (chips).

IEC 61499 allows function blocks that encapsulate software functionality and algorithms to be defined in a standard format. This allows tools and other standards that deal with function blocks to use the same concepts and methodology. The IEC 61499 standard also defines a range of communication function blocks, such as Client/Server function blocks, which can be used to formalise the exchange of data between function blocks in different physical processing resources. There are also service interface function blocks to provide interfaces with the processing resource infrastructure.

Figure 1.5 shows three interconnected function blocks, representing the connections between a pressure transmitter and PID control function block and a pump using concepts from the IEC 61499. Notice that there are both data and event flows between function blocks. We will see later that the IEC 61499 methodology allows data and its associated event to be closely coupled, that is, to be coherent or alternatively for events and data to be handled asynchronously.

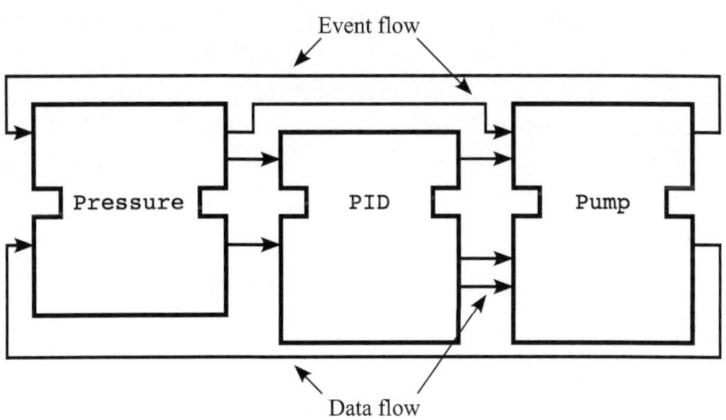

Figure 1.5 Function block data and event flows

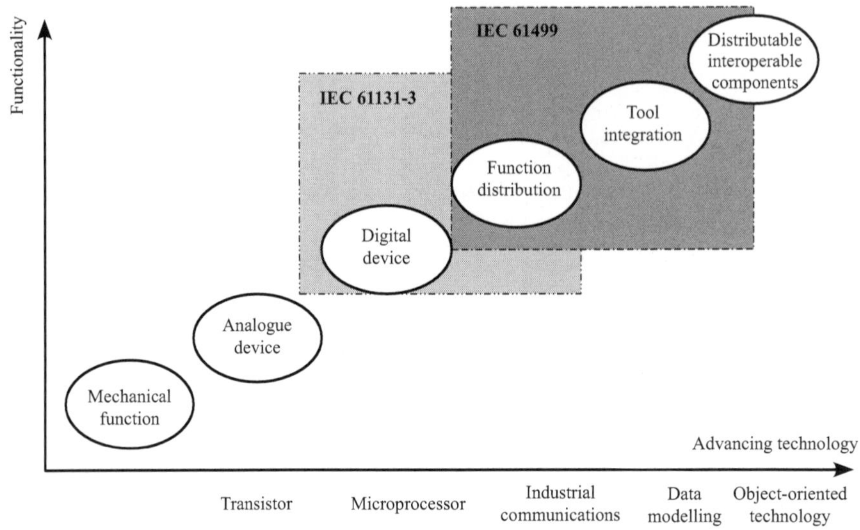

Figure 1.6 Development of technologies for industrial control

Figure 1.6 depicts the trends in industrial control technology during the last 50 years; since the 1950s, there has been a steady growth in the functionality provided by control systems due to advances in both hardware and software. As control systems became digital, using microprocessors, there has been an increased need for standards to reduce unnecessary diversity in software and lessen cross-system integration problems.

IEC 61131-3 has focused on standardising PLC languages for single processors or small configurations with a few closely coupled multiprocessors. With the move to large scale distributed functionality, there is a need for further standards such as IEC 61499 to harmonise the way functionality is defined and distributed. There is also a growing requirement that all the related system build tools can be integrated as well.

For example, all the software tools used to design, configure and manage a distributed system should run as an integrated suite. The design tool that defines a system should be integrated with tools for programming and configuring devices along with tools for defining HMI screens and configuring industrial networks. It is the intention that IEC 61499 will define system models that will help not only with the design of functionality in distributed systems but also with the integration of system tools through the definition of data and information models.

1.2 Development of function block concept beyond IEC 61131-3

Why was it not possible to use the function block concepts defined in IEC 61131-3 for distributed systems? There are a number of limitations with the original function block concept introduced by the PLC Languages Standard IEC 61131-3. With the IEC 61131-3 Function Block Diagram (FBD) graphical language, function

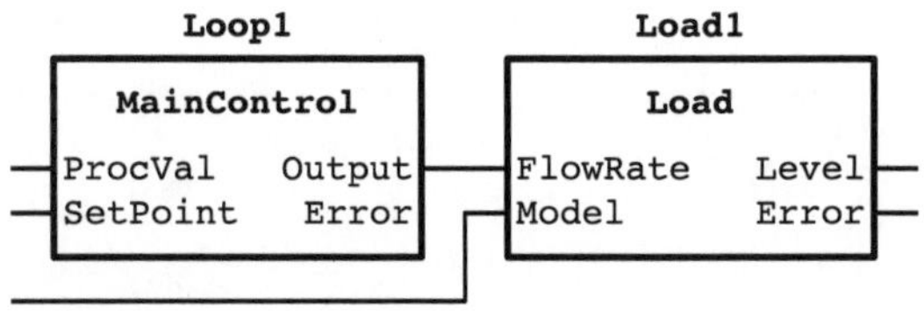

(a) Function block network

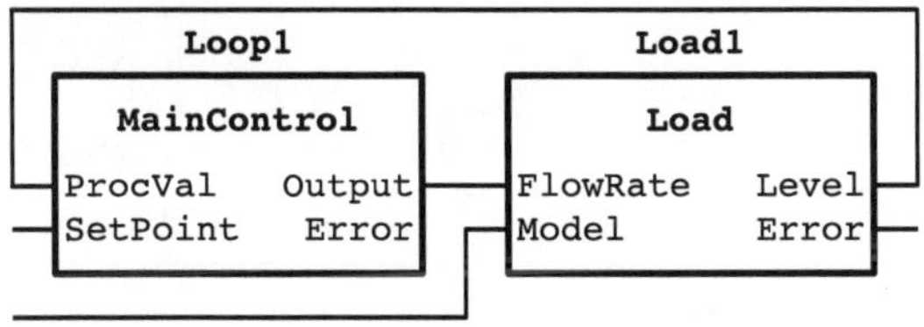

(b) Network with feedback connection

Figure 1.7 Using IEC 61131-3 function blocks

blocks can be linked by simply connecting data flow connections between function block input and output variables, see Figure 1.7a. Each function block provides a single internal algorithm that is executed when the function block is invoked. The normal execution order is determined by the function blocks dependency on the other function blocks; the order normally runs from left to right because function blocks to the right depend on the output values of function blocks on the left.

However, when a feedback path is introduced, as shown in Figure 1.7b, the execution order cannot be determined from the diagram, since the execution of both function blocks depends on an output value of the other function block. In a complex network, it is very difficult for a programming system to determine a valid order of execution. To overcome this problem, many IEC 61131-3 programming systems provide additional mechanisms to define the execution order of function blocks. For example, the user can view a list of function blocks and manually assign an execution order. Unfortunately, such mechanisms are outside the scope of the IEC 61131-3 standard. As a consequence, an important aspect of a function block network is that the method for defining the execution order of function blocks is not consistent or portable across different control systems.

There is one feature in IEC 61131-3 that does provide a crude mechanism for passing execution flow through a chain of function blocks that is worth considering: the use of the EN input and ENO output signals (Figure 1.8). The EN and ENO

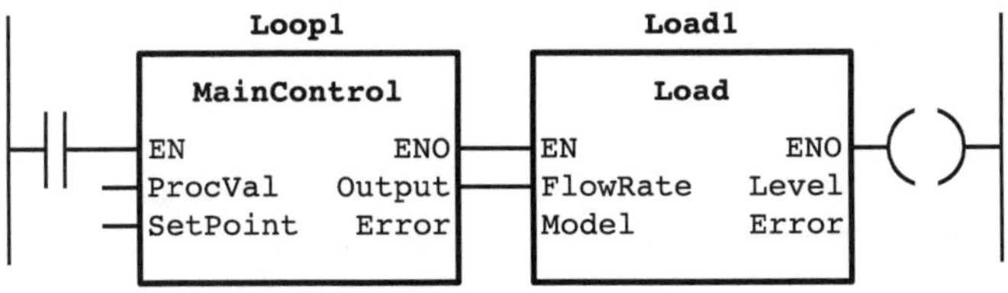

Figure 1.8 Execution control using IEC 61131-3 EN and ENO

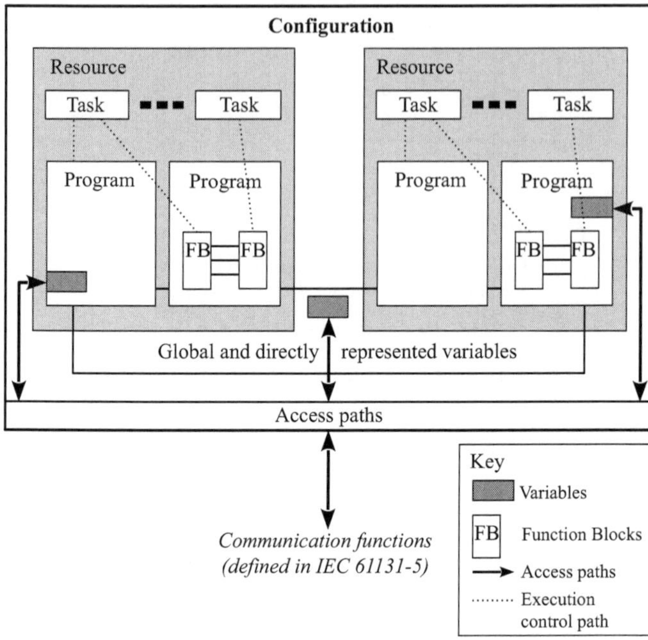

Figure 1.9 IEC 61131-3 software model

signals were intended for function blocks to pass 'power flow' when used in rungs of a Ladder Diagram (Figure 1.8). However, it is now recognised that the EN and ENO do not provide the degree of flexibility needed for complex function block networks. In effect, the EN and ENO signals can be regarded as a means to pass events between function blocks. EN signals that the function block may be invoked because its input data is ready; ENO is signalling that the function block has executed and the output data is ready for the next function block. We will see that this idea of event passing has been extended in IEC 61499.

The focus of the IEC 61131-3 standard has been to define a software model and languages for PLCs where software is typically running on one or a few tightly coupled processing resources. Therefore, the IEC 61131-3 software model, see Figure 1.9, does consider configurations that have multiple resources. The standard provides two different mechanisms for passing data and control signal between resources, namely global variables and communications function blocks.

1.2.1 Global variables

Using global variables located at the configuration level, it is possible to transfer data and control signals between programs and function blocks running in different resources. However, it is well understood that the use of global variables is a very poor and sometimes unsafe mechanism for handling data transfer between procedures running in different processors. It is not possible to clearly identify where global variables are updated and where they are used. There is no graphical means in IEC 61131-3 to define the linkage between global variables and the variables referencing them, that are located inside programs and function blocks. There are also more serious problems with global variables because the timing and synchronisation of signals passed by global variables is difficult to define. Furthermore, the mechanism provided within the configuration for handling the initialisation and updating shared global variables are not defined in IEC 61131-3.

1.2.2 Communications function blocks

Part 5 of the PLC standard, IEC 61131-5 [13], is concerned with communications services for PLCs programmed using the IEC 61131-3 software model. IEC 61131-5 defines a range of function blocks that can be used to exchange data between PLCs. This includes function blocks to allow a PLC to function as a 'server', i.e. allow a PLC to support and respond to external service requests. There are also function blocks to support 'client' behaviour i.e. support services that enable a PLC to request and control another PLC or system functioning as a server.

IEC 61131-3 allows a subset of variables within a configuration to be accessed externally. These are called ACCESS variables and can be accessed via a communications interface from an external PLC using communication function blocks or can be accessed from other non IEC 61131-3 devices using services that are outside the IEC 61131-3 standard.

We have seen that IEC 61131-3, along with the manufacturing messaging standard IEC 61131-5, provides a range of different software mechanisms for allowing PLCs to communicate. These are quite adequate for systems with only a few PLCs. However, it was clear to the IEC working group developing the function block standard that the IEC 61131 communications model has serious limitations. Concepts such as global variables and communications function blocks do not provide a clear and concise method for defining the connections between distributed function blocks.

A consistent communications model was required that could be used not only for PLC to PLC communications but also between devices large and small distributed over industrial networks. The new function block model had to be scalable and extensible – so that it would be equally applicable to modelling the communications between control systems, PLCs and controllers as between smaller Fieldbus devices, such as smart valves and sensors. In fact, a function block model was sought that would cover all types of devices and controllers.

To summarise, the main deficiencies of the software model provided by IEC 61131-3 for distributed systems are as follows:

- Applications in the IEC 61131-3 model are not distributable over multiple resources.
- The function block execution order is not always clearly defined and not under the direct control of the application developer.
- The assignment of tasks to programs and function blocks does not provide sufficient flexibility.
- The 'scanned' nature of IEC 61131-3 function block execution cannot be mapped to function blocks connected across distributed resources.

1.3 Development of IEC 61499

When the International Electrotechnical Commission first identified the need for a new function block standard for distributed industrial process measurement and control systems in 1990, it was realised that these function blocks would be a generic concept that could be applied to a wide range of standards. For example, function blocks concepts can be used within PLCs, smart devices, building management systems and Fieldbus protocols. At that time there was already a standard utilising a function block concept under development, namely IEC 61131-3 focusing on the harmonisation of PLC programming languages. In order to leverage the existing know-how, the IEC technical committee TC 65 assigned the work on IEC 61499 to the same working group (at that time WG 6). The IEC working group for IEC 61499 has members from the United States, Japan, the United Kingdom and many European countries, who represent both industrial control system suppliers and end-users.

IEC 61499 is a multi-part standard that took a number of years to complete. The IEC standardisation process is managed in four main phases: (i) the standard development phase, (ii) the Publicly Available Specification (PAS) phase, (iii) the final review and publishing phase and finally (iv) the maintenance and republishing phase. The last phase is performed every 5 years. PAS is a concept being promoted by the IEC for publishing standards early, avoiding some of the more time-consuming bureaucratic stages of standards approval. A PAS can be published even though the standard has not completed all the international standardisation approval processes. A PAS can be regarded as a trial standard that is made available to the industrial community to allow the early development of products and services.

For the first part of IEC 61499, the PAS phase started in 2000 and resulted in the final publication of all parts as an international standard in 2005. Feedback gathered during the PAS phase resulted in several changes and improvements making IEC 61499 easier to implement and apply. In 2010 the first maintenance was performed. During this first maintenance phase, several issues and ambiguities, which had been identified since the publication of the first edition were corrected. The resulting second edition of IEC 61499 was published in 2012. The second edition is also the basis of this book and consists of three parts: 1, 2 and 4[4].

Part 1 covers the architecture and concepts for designing and modelling distributed function block–oriented systems and covers the following topics:

1. General requirements, including an introduction, scope and normative references (i.e. to other standards), definitions and reference models.
2. Rules for the declaration of function block types, and rules for the behaviour of instances of function block types.

[4] Part 3 has been withdrawn in 2008 (see Section 8.1 for more details).

3. Rules for the use of function blocks in the configuration of distributed industrial process measurement and control systems (IPMCS).
4. Rules for the use of function blocks in meeting the communication requirements of distributed IPMCS.
5. Rules for the use of function blocks in the management of applications, resources and devices in distributed IPMCS.

The main focus of this book concerns the architecture and models defined in Part 1 of the IEC 61499 standard.

Part 2 of the IEC 61499 standard [14] addresses the definition of formal information models that will enable CASE[5] tools and utilities to manipulate and exchange system designs based on function blocks. The main focus of Part 2 is the 'engineering task support' and it provides guidance on engineering tasks concerned with the design, implementation, operation and maintenance of IPMCS constructed using the architecture and concepts defined in Part 1.

Originally the STandard for the Exchange of Product data, STEP (ISO 10303), has been considered as the means by which function block designs are saved and exchanged across different design workstations. STEP is used by CAD stations for storage and exchanging of electronic circuit designs in an implementation independent form. There is clearly a strong synergy between electronic schematics and control system designs based on function blocks.

However, finally the eXtensible Markup Language (XML) has been selected in IEC 61499 Part 2 as a means to save and exchange function block definitions. This provides the exciting possibility of being able to transfer designs across the Internet and view them using Web browsers. XML has been developed as major enhancement to the Hypertext Markup Language (HTML) currently used for Web page creation. Using XML will allow designs to be saved with various attributes including version information and graphical layout details – this is discussed further in Section 2.6.

Part 4 [15] targets the important topic of how compliance to IEC 61499 is defined and how vendors need to specify it. Already very early in the development compliance was identified as a very important consideration particularly with distributed systems where devices from different vendors are used in a system. On the other hand (as stated) IEC 61499 is generic and therefore does not define a concrete implementation of a distributed IPMCS in all details. In order to address this issue, Part 4 of IEC 61499 defines rules for structuring and developing compliance profiles. A compliance profile as defined in IEC 61499 Part 4 needs to specify the implementation of the features of IEC 61499 Part 1 and Part 2 considering the important properties of distributed IPMCS:

- *Interoperability* provisions define which means are used allowing devices from multiple vendors to interact via communication systems.

[5] CASE – Computer Aided Software Engineering.

- *Portability* provisions define an exchange format for IEC 61499 entities (e.g. function blocks, applications, system configurations) allowing software tools from multiple vendors to produce, parse and interpret them correctly.
- *Configurability* provisions define means that allow software tools from multiple vendors to configure devices from multiple vendors.

It was the original intention of the IEC that the new function block standard would become a generic standard that could be used as a basis for standards throughout the domain of industrial process measurement and control. For this reason, because of its generic nature, IEC 61499 appears to be a rather academic standard. It has been deliberately defined to be 'application domain neutral', i.e. it contains no specific features for any particular industrial application area. It is designed so that other standards can build on the IEC 61499 concepts and add their own domain specific extensions.

A good example of a standard built on the IEC 61499 function block model is demonstrated by the Process Control Function Block working group. This group has the primary objective of defining function blocks for use in the process industries, but they have taken concepts from the generic function block model in IEC 61499 as the basis of their work. By applying the generic model to real industrial process control applications, the process control group has provided useful feedback to the IEC 61499 working group. In many cases they have highlighted shortcomings in the function block model that have resulted in improvements to IEC 61499.

Furthermore, during the evaluation phase of the standard, it was found that the function block models as defined in IEC 61499 had great potential for improving system design. In the past few years several platforms (i.e. engineering tools and control devices) implementing the concepts of IEC 61499 have become available. These allow the development and operation of distributed IPMCS directly using the models of IEC 61499.

1.4 Why use function blocks

To many software engineers, the idea of function blocks seems to some degree archaic – a strange software paradigm that appears to represent software as pieces of hardware. In effect, that is exactly what a function block is, a model of software that treats the encapsulated behaviour in a form that is similar to an electronic circuit. However, in improving the software development process, the key aspect has always been to add further abstractions and higher level software modelling aspects in order to cope with an increased complexity.

Objects, used in the Object-Oriented (OO) software world which are in some respect similar to function blocks, have become successful because they can be used to model the behaviour of entities and concepts in the real world. The main benefits from using objects can be summarised as:

Objects reflect the real world: When designing an application, it is more natural and intuitive to represent real-world entities associated with an application as objects, e.g. document, employee, and product.

Objects are stable: Generally, objects are proven software elements that do not change significantly. In many cases, developers use the same object classes in a wide range of applications. For example, when an object is created that represents all the behaviour and characteristics of an entity such as 'product supplier', it can be used in a wide range of different business applications dealing with suppliers. A 'product supplier' object would typically have details such as name, address, product ranges, trading terms etc., and methods to obtain and update this information.

Objects reduce complexity: A developer can work with an object without really understanding how the object works internally. An application can be developed by creating and linking objects – there is generally no need to understand the object's internals.

Objects are reusable: Once an object is developed and tested it can become part of a developer's repertoire. In some cases, an object can be published in a library, where it can be used by developers either locally or even globally.

While OO software development greatly changed the world of software development, it could not keep up with the promise of providing extensible reusable software. It has been shown that objects are very often designed to be too specific for their use-case or depend on the context in which they are used, limiting their general applicability [16].

In order to improve this situation the stronger concept of *software components* has been introduced by Szyperski [17]. He defines a software component similar to an object: "A software component is a unit of composition with contractually specified interface and explicit context dependencies only. A software component can be deployed independently and is subject to composition by third parties." [17, Section 4.1.5]. Software components have the following main properties:

- *Unit of independent deployment*: This property requires that a component is self-contained and is clearly separated from the environment and other components. A component will never be deployed partially.
- *Unit of third-party composition*: Third parties who have no information on the internal construction of components can use them in combination with other components.
- *Component interaction only via explicit interfaces*: In order to provide the first two properties the interaction with components has to occur via clearly defined interfaces. A purely static interface definition (e.g. method signature) is not enough; dynamic interface definitions on how to correctly use it are also required. Furthermore, components must not have hidden interfaces (e.g. global variables).

As in component-oriented software development, in the function block world, the system designer's main focus is to take standard, proven encapsulated functionality and link it together in the quickest and most intuitive way possible. The use of function blocks is nearer to the mind-set of the industrial system designer who is familiar with connecting physical devices together in different ways to provide a particular system solution. Function blocks also share most of the characteristics of software components, which results in some significant benefits to the system developer and end-user:

- The quantity of control software to be developed for an application is reduced by using function blocks.
- The time required to develop control systems is reduced.
- Control systems using the same types of function block will have more consistent behaviour.
- The quality of control systems will be improved.

The current trend in software development is to use abstract formal models for describing software at an even higher level [18]. Especially by using domain specific modelling approaches certain software systems can be developed by domain experts with less effort at higher quality. As we will see in the following chapters, IEC 61499 provides both an effective means for software modelling and a comprehensive modelling language that addresses the design of distributed control systems, notably in the IPMCS domain.

1.5 System design views

The design of software for any large project can be very complex. Especially where there is also some aspect of distributed control involving software running in different processing resources, the design problems can be daunting. There is a clear requirement for a number of graphical design views to allow the different aspects of a design to be analysed and expressed. Some views will express the abstract aspects of the design, while others are required to show the physical structure of the system or the way the software is organised.

No matter how hard people have tried, it is just not possible to convey all aspects of a system design using one design methodology. There are so many design issues to consider that they cannot be expressed in a single type of graphical notation, such as:

- What is the top-level software structure?
- What functionality does the system deliver to its end-users?
- How is the functionality distributed throughout the system?
- How are the system components connected?
- How are the software libraries and standard components managed?
- How does the system respond to certain critical events?

Many system design problems are a result from trying to use too few or inappropriate different design methodologies to depict all the aspects of a system design. A particular design view might be able to show how software for a system is logically connected, but it would not be able to depict the way the system responds to events.

In fact, it is now recognised that most complex software designs can require at least four different design views and a set of scenarios – this forms an

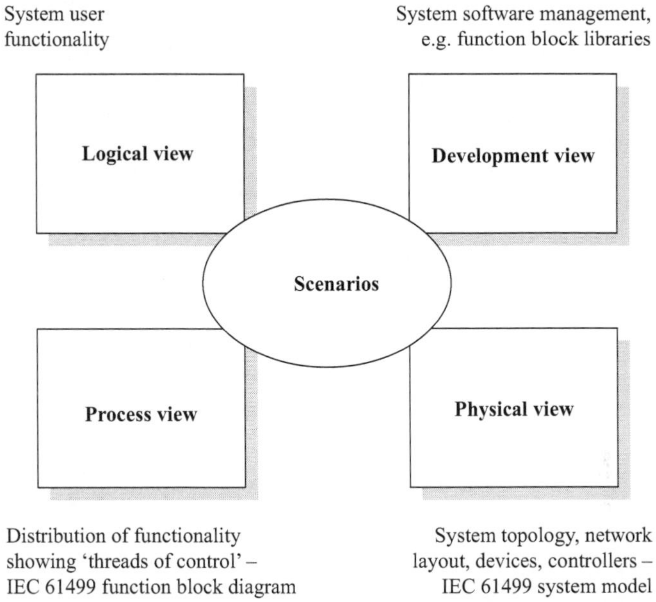

Figure 1.10 The 4 + 1 View Model of system development

architecture known as the 4 + 1 View Model, as proposed by Kruchten [19] (Figure 1.10).

Although Kruchten has considered applying these design views to the world of object oriented software, the same design views are also applicable to distributed control system design.

1.5.1 Logical view

This design view is used to depict the functional requirements of the system. It expresses the software functionality as required by the system user. In a distributed system design, it would show the main software function blocks and the main interfaces between them. Issues such as how the system functionality is distributed and executed are not addressed.

A methodology such as the IEC 61499 applications as well as the function block model itself can be used to define some aspects of the logical design view.

1.5.2 Process view

The process design view is concerned with many of the non-functional requirements of a system; these include performance, system distribution and issues such as concurrency. Kruchten defines the process view as depicting 'logical networks of communicating programs that are distributed across a set of hardware resources'.

This corresponds almost exactly to the concepts in the IEC 61499 Function Block standard, which provides an architecture for depicting the implementation view of a distributed system as networks of interconnected function blocks.

1.5.3 Development view

The development view depicts how the software that is used to build a large system is organised. Building a large distributed control system will involve numerous software libraries and software modules. The development view shows the relationships between software components, such as function blocks in terms of ease of reuse, constraints, component size and version compatibility. For example, consider a function block used for conveyor control, where we would want to indicate which device types support it and we would like to show any constraints on other function blocks that need to interact with it.

Currently there is no IEC standard methodology that deals with the development view of a distributed control system. With the interface concept provided by the adapter model, in IEC 61499 only the limited aspect of the provided and required interaction points can be defined.

1.5.4 Physical view

In a distributed control system, the physical view is well understood. It depicts the physical devices and controllers in a system and shows the various network communications links between them. A physical view will generally consider the physical configuration of the system showing the location of devices and details on bus and communications links.

This corresponds partly with how IEC 61499 specifies the system model, which provides means to model available devices, communication links and the interconnection of the devices. Physical placement in the plant as well as interaction with the process is not considered.

1.5.5 Scenarios

The last but important design view that completes any system design is what Kruchten calls *scenarios*. A scenario depicts the major interactions between units of software to provide the most important, key functionality of a system. For example, in a distributed control system, some important scenarios to consider might be system start-up, device fault detection and recovery, recipe activation and system shutdown. Each scenario would consider the interactions between the different functional parts of the software. A scenario might show both aspects of the logical and process design views.

By describing the various scenarios, the designer can review and test the design by asking a series of 'what if?' questions. The design cannot be considered to be complete until all the key scenarios have been defined. There is currently no methodology defined by any IEC standard that can be used to define scenarios for distributed control systems.

From this quick overview of the 4 + 1 View Model of Architecture, it is clear that IEC 61499 provides at least in parts three of the five design views required for distributed control systems. However, IEC 61499 does represent an important step towards a unified design architecture. The other views will no doubt emerge as designers start to face the challenge of building large distributed systems.

1.6 The future beyond IEC 61499

The function block model proposed by IEC 61499 has been criticised for not adopting concepts from OO software technology as has been done with the latest revision of IEC 61131-3. However, in addition to OO software technologies, especially higher level concepts like programming in the large and extended software modelling concepts would be needed. The standard has started by modelling existing industrial function block concepts but extensions to move towards these concepts will undoubtedly need to be considered in the near future.

New industrial standards for communications and software components will clearly bring benefits in allowing physical devices and software to be readily interconnected. However, before we can achieve truly interoperable software components that can be used to implement large systems, we need to agree on general methods for describing requirements such as information models and data transformations. It is the intention that IEC 61499 should be able to address this problem in the domain of industrial control systems.

In the following chapters we will review the concepts from IEC 61499 and see how this standard can be used to model the design of distributed control systems.

Chapter 2

IEC 61499 models and concepts

We will now review the main models and concepts defined in IEC 61499 to gain a general overview of the function block standard. It is advisable to have some understanding of the material in this chapter before proceeding into any of the following chapters where we will review specific features of IEC 61499 in more detail.

Topics covered in this chapter include:

- *characteristics of function blocks and their execution,*
- *the different forms of function blocks,*
- *service interface function blocks to provide interfaces into hardware and operating systems,*
- *models to represent applications independent from the system configuration,*
- *the system, device and resource models for distributed control systems,*
- *the distribution model which assigns applications to the devices and resources,*
- *exchange and storage formats for IEC 61499 entities.*

Before we proceed to look at the many models and concepts introduced by IEC 61499 in detail, let us reconsider the scope of the IEC 61499 standard as first discussed in the introductory chapter. Surprisingly the primary purpose of IEC 61499 is not as a programming methodology but as an architecture and model for distributed systems. IEC 61499 provides a set of models for describing the behaviour and structure of distributed industrial process measurement and control systems using the function block concept. This is an important distinction and must be understood to avoid many of the misunderstandings that exist about IEC 61499.

IEC 61499 provides terminology, models and concepts to allow the implementation of a function block oriented distributed control system to be described in an unambiguous and formal manner. Having a formal and standard approach to describing systems will allow systems to be validated, compared and understood. This is the first step towards standard programming methodologies for distributed systems. The IEC 61499 standard writers have taken the view that it is not possible to have a consistent programming methodology unless there is a consistent architecture that underpins what we are trying to program.

We will now review the various models introduced by IEC 61499 that together form the architecture for a function block oriented distributed industrial process measurement and control system.

2.1 Function block model

At the core of the standard is the **function block** model that underpins the whole IEC 61499 architecture. A function block is described as a 'functional unit of software' that has its own data structure which can be manipulated by one or more algorithms. A **function block type** definition provides a formal description of the data structure and the algorithms to be applied to the data that exists within the various instances.

This is not a new concept but based on common industrial practice applied to reusable control function blocks of various forms. A good example is the Proportional, Integral and Derivative (PID) function block used in many PLCs and controllers. The system vendor will typically supply a type definition for a PID function block. The programmer can then create multiple instances of the PID function block within the control program, each of which can be run independently. Each PID instance, such as 'Loop1', 'Loop2' will have its own set of initialisation parameters and internal state variables and yet share the same update algorithm.

2.1.1 General characteristics

IEC 61499 defines several forms of function blocks, which we will review in detail in later chapters. The main features of an IEC 61499 function block are summarised as follows:

- Each function block has a type name and an instance name. These should always be shown when the function block is depicted graphically.
- Each function block can have event inputs, which can receive events from other function blocks via event connections.
- Each function block can have event outputs, which can be used to propagate events on to other function blocks.
- Each function block can have data inputs that allow data values to be passed in from other function blocks.
- Each function block can have data outputs to pass data values produced within the function block out to other function blocks.
- Events can be associated to data using the `WITH` qualifier. In the graphical representation this is shown using a small square connector that links the event with its associated data. For inputs and outputs the meaning is as follows:
  - On the occurrence of an input event its associated input variables are updated (i.e. sampled from the connection), other input variables retain their values (i.e. are unchanged).
  - On the triggering of event outputs the associated data outputs will be made available (e.g. sampled onto the connection), other outputs stay unchanged.
- The IEC 61131-3 data types are utilised for the variables' data types (see Appendix A for an overview on available data types).
- Function blocks encapsulate functionality which may contain internal variables. The kind and form of the encapsulated functionality depends on the function block type.

- As IEC 61499 does not allow global variables, the encapsulated functionality only has access to function block input, output and internal variables.
- All function block data (i.e. input, output and internal) is retained between function block invocations.

In Figure 2.1, the main characteristics of an IEC 61499 function block are depicted. The top part of the function block, called the 'Execution Control' portion, contains a definition to map events on to encapsulated functionality. That is, it defines which encapsulated functionality defined in the lower body is triggered on the arrival of events at the 'Execution Control' and when output events are triggered – what the standard calls the 'causal relationship among event inputs, event outputs and the execution of encapsulated functionality'. The standard defines means to map the relationships between events arriving at the event inputs, the execution of encapsulated functionality and the triggering of output events depending on the type of function block – this will be discussed in later chapters on the different function block types.

The lower portion of the function block contains the encapsulated functionality with possible internal data – both of which are hidden within the function block. A function block is a type of software component and, if well designed, there should be no requirement for a user to have a detailed understanding of its internal design.

A function block relies on the support of its containing resource to provide facilities to schedule encapsulated functionality and map requests to communications and process interfaces.

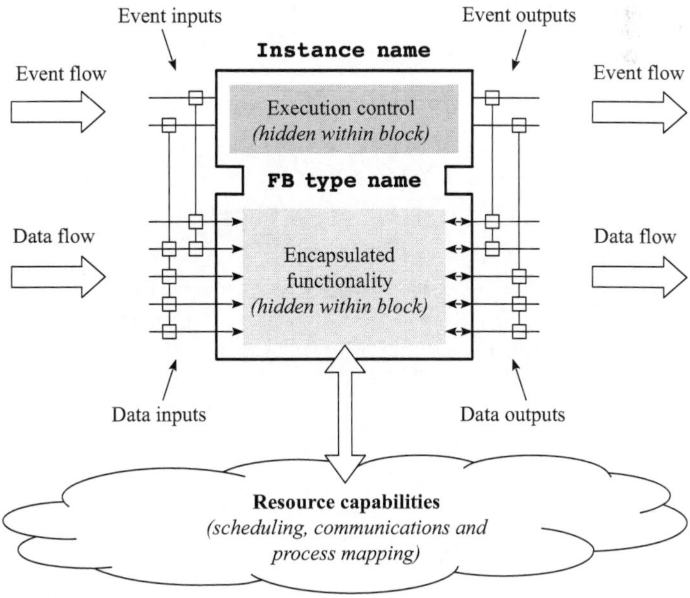

Figure 2.1 Function block characteristics

The standard states that a resource may optionally provide additional features to allow the internals of a function block to be accessed. Clearly, say, in a Fieldbus device, it would be always useful for maintenance or commissioning purposes to be able to examine the internal variables within a function block. So there may be 'back-door' methods to access function block internals; however, from the IEC 61499 architecture view point, control variables and events are only passed by the external exposed interfaces.

Note: IEC 61499 function blocks contain all algorithms and initialisation values to define their complete behaviour.

2.1.2 Execution model for function blocks

The execution model generally defines function blocks as passive elements. They require a trigger by an input event in order to invoke its encapsulated functionality. The execution order is best described with the aid of Figure 2.2. The numbered features on the function block show the order in which the different parts of the function block are executed. The model assumes that the execution environment[1] in which a function block exists provides a scheduling function that ensures that each phase of function block execution occurs in the correct order and at the correct priority. Furthermore the standard defines the execution of a function block as algorithmic. That means that it terminates in a finite time. Therefore, blocking elements may not be used within a function block. For real-time constrained applications the execution time constraints can be stronger (i.e. below an application specific threshold).

There are a number of discrete phases, each of which may take some period of time to elapse, required for the function block to execute; each phase depends on

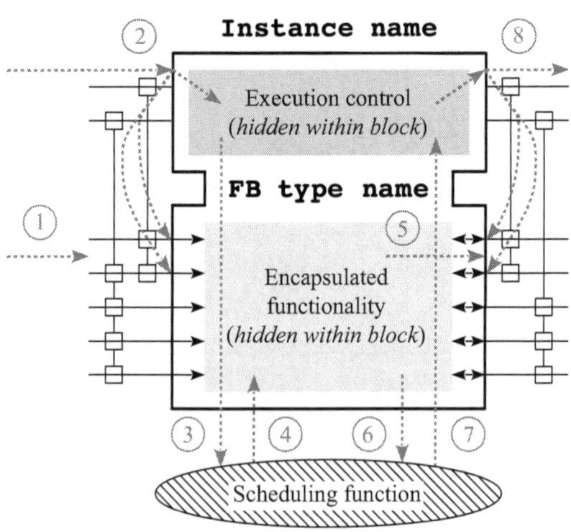

Figure 2.2 Execution model for function block

[1] In standard terminology a resource, see Section 2.3.2.

defined interactions between the function block and the underlying scheduling function. Figure 2.2 depicts the eight steps that must occur sequentially for the function block to operate; the termination of each phase is defined by a particular numbered step.

① Data values coming from other function blocks are made available at the function block data inputs.
② An input event arrives resulting in the sampling of the associated input variables (i.e. `WITH` qualifier) and in the notification of the execution control.
③ Based on the function block's current state the execution control signals the scheduling function that a certain element of its encapsulated functionality is ready to execute.
④ After some period of time as determined by the loading and performance characteristics of the resource, the scheduling function starts to execute the requested function block's encapsulated functionality.
⑤ The encapsulated functionality performs its task, processes input values and, in some cases, also processes internally stored values to create new output values that are written to the function block's outputs.
⑥ The encapsulated functionality completes its execution, and signals this to the scheduling function to indicate that updated output values are stable and ready.
⑦ The scheduling function invokes the function block's execution control, notifying it about the finished execution of the encapsulated functionality and enabling it to generate an output event. Different output events (or also none) may be generated depending on which input events have arrived and the internal state of the execution control.
⑧ The execution control in turn creates an appropriate output event at the function block's output event interface, and associated output variables are made available to the connected function blocks (e.g. sampled onto the connection). The output event is used to trigger the execution of downstream function blocks, signalling that they can now use output values generated by this function block.

Note: After the input variable sampling in step ②, the input variables are not changing during the whole function block execution triggered by an input event occurrence. This ensures stable and deterministic data values during the execution of encapsulated functionality.

The presented sequence of execution steps is the most basic execution sequence that occurs as a result of an input event arrival. Depending on which input events have arrived and the internal state of the execution control the steps ③–⑧ can be taken several times. In such cases several output events may be sent in response to the triggering of one input event.

It is important to note that there are a number of constraints on this execution model. These timing phases cannot overlap and must occur in the prescribed order for the function block to execute correctly. However, in some implementations, some phases can be so short in duration as to be regarded as being instantaneous. IEC 61499 does not define limits on any of these times. However, it does state that in any function block model it should be possible to define the duration of these different phases in order to accurately model the timing characteristics of the complete function block network.

The standard defines the following durations that will be significant when building applications:

$\mathbf{T_{setup}} = T_{②} - T_{①}$ Time between the availability of input values (i.e. updated by preceding function blocks) and the arrival of an input event triggering their processing.

$\mathbf{T_{start}} = T_{④} - T_{②}$ Time between receiving an input event and executing the function block's encapsulated functionality. This duration may depend on the resource loading, i.e. how many other function blocks are also in the scheduling function's pending queue.

$\mathbf{T_{algorithm}} = T_{⑥} - T_{④}$ Time between starting and completing the function block's encapsulated functionality.

$\mathbf{T_{finish}} = T_{⑧} - T_{⑥}$ Time from finishing the encapsulated functionality and triggering the output event.

The relationship between these timing points are depicted in Figure 2.3; the points where the input and output data from the function block change are also shown. It should be noted that the standard assumes that events behave as discrete points in time and have no duration. In a physical system, events may require the transfer of some form of state change information between blocks and may have a short but finite duration.

The IEC 61499 model assumes that there are no input event and data queues associated to a function block's inputs. However, the standard defines that the execution environment needs to assure that only one input event is delivered at any

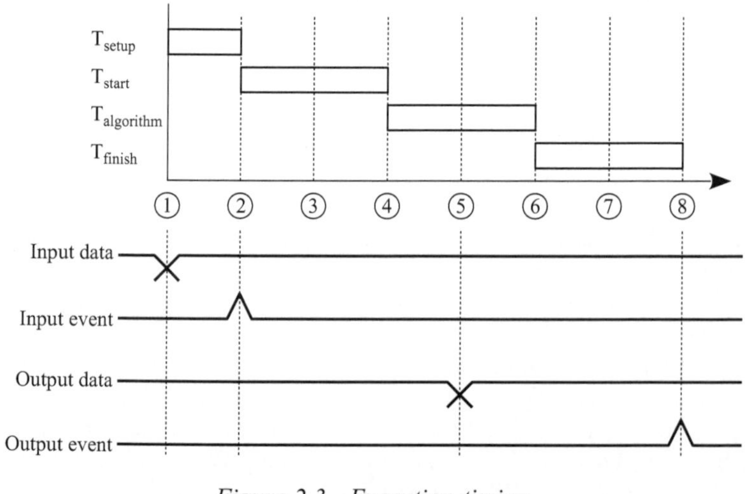

Figure 2.3 Execution timing

given instant in time. Execution environments can optionally queue events in order to not lose events during function block execution. It is very important for the underlying execution environment to detect when a function block receives input events at a faster rate than can be handled by the function block. Furthermore, an implementation may provide some form of input data queuing, but this is not explicitly modelled by IEC 61499. However, such behaviour may be modelled by creating special function blocks. The model does allow resources to provide facilities to process function block algorithms using multitasking. With appropriate prioritisation of events within the scheduling function, it is possible for most event overloading problems to be avoided. However, note that the characteristics and internal behaviour of the scheduling function are not within the scope of this standard. A detailed description of different implementation options for executing IEC 61499 function blocks is given in [20].

2.1.3 Function block types

An important concept in IEC 61499 is the ability to define a function block type that defines the behaviour and interfaces of function block instances that can be created from the type. This is synonymous with how in object-oriented software the behaviour of object instances is defined by the associated object's class definition.

A function block type is defined by a type name, formal definitions for the function block's input and output events, and definitions for the input and output variables. The type definition also includes the internal behaviour of the block but this is defined in different ways for different types of function blocks.

Basic function block types

The behaviour of a basic function block is defined in terms of algorithms that are invoked in response to input events. As algorithms execute, they trigger output events to signal that certain state changes have occurred within the function block. The mapping of events on to algorithms is expressed using a special state transition notation called an Execution Control Chart (ECC).

Composite function block types

The internal behaviour of composite function blocks is defined by a network of function block instances. The definition therefore includes data and event connections that need to exist between the internal function block instances.

Service interface function block types

Service interface function blocks provide an interface between the function block domain and external services, for example, to communicate with function blocks in a remote device or to read the value of a hardware real-time clock. Because a service interface function block type is primarily concerned with data transactions, it is defined using service sequence diagrams. Service sequence diagrams are described in more detail in Section 3.3.

2.2 Application model

An IEC 61499 **application** is defined as a network of interconnected function blocks, linked by event and data flows. An application therefore consists of function block instances and interconnection definitions which, in some cases, include multiple instances of function blocks of particular function block types. It is a principle of IEC 61499 that all behaviour is defined in terms of function blocks. As a result, we will see that there are no global or local variables in an application that can exist outside of function blocks. This is an important distinction between an application program created for a PLC based on IEC 61131-3 and an IEC 61499 application.

The features of the IEC 61499 application model are shown in Figure 2.4[2]. Event connections are in general always point to point connections, i.e. from an event output to an event input[3]. For data connections the standard allows a fan-out connection style. That means that one data output can be connected to several data inputs. A fan-in, several data outputs connected to one data input, is not allowed. The reason for this is that function blocks are fully decoupled in IEC 61499 and have no information about the source of input connections. Therefore, if a fan-in were allowed, a function block could not determine from which connection to take the data. Finally, incoming event and data connections do not need to originate from the same function block (e.g. the inputs of `FB1` in Figure 2.4). This is, for example, a helpful feature if the sampling and processing of data that originates from several sources needs to be synchronised with the occurrence of one event.

Note: For brevity the names of inputs and outputs (events and data) have been omitted in Figure 2.4.

In practical terms, an application is the entire set of function blocks and interconnections to solve a particular automation control problem. Examples might be the set of function blocks required to control a production line, a plastics extruder or a fermentation vessel. However, the application model at this stage does not consider any particular hardware and focuses only on the overall control

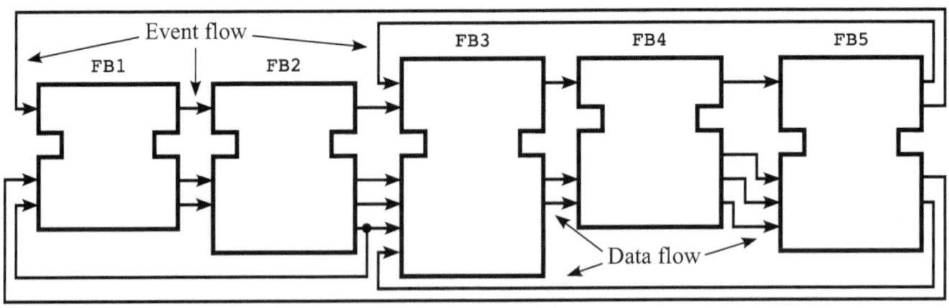

Figure 2.4 Application model

[2] Arrow heads on connectors for data and events are not mandated by the standard but can be used to aid understanding of event and data flows.

[3] See Section 6.3.1 for a detailed discussion on event multiplexing and demultiplexing.

functionality. Therefore, in contrast to IEC 61131-3 application development which is very resource focused, the IEC 61499 application development is application centred as the application is the core element in the system development. This has the great advantage that the control engineer can focus on the functionality. This at first reduces the impact of late hardware changes and also fosters the reuse of general control functions. Furthermore, engineering tools can provide facilities to test and simulate applications independent of the specific control hardware, allowing control engineers to evaluate the functionality at early stages in the development process.

2.3 System model

At the physical level, a distributed system consists of a set of devices interconnected by various networks to support a set of co-operating applications. An application, such as the control of a production line, process vessel or conveyor will typically require the interoperation of software running in a number of devices. Until very recently, a typical distributed application has involved a small percentage of software running in remote devices such as loop and temperature controllers, while still having the main intelligence back in a central device such as a PLC. As devices such as smart sensors and actuators start to provide more processing capability, software functionality can become truly distributed across many more devices, to a point where it is difficult to identify a main controlling device.

2.3.1 Overall system structure

We will start by looking at the top level **system model** defined in IEC 61499. This model defines available control devices and the communication relationships between these, forming a network of communicating devices. Communication links can have different types and devices may be connected to different communication segments. This allows the formation of communication hierarchies as shown in Figure 2.5.

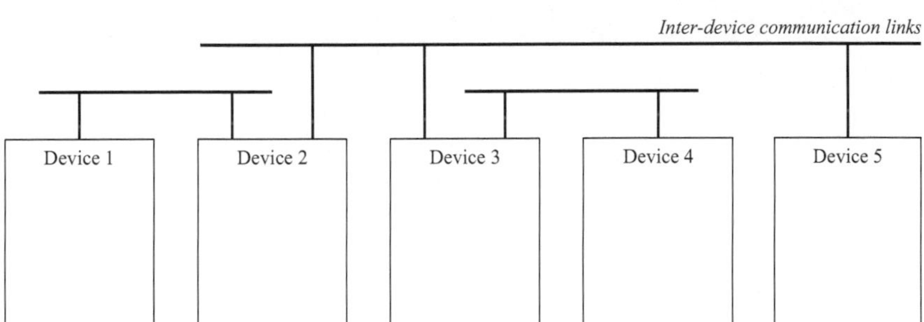

Figure 2.5 System model

2.3.2 Device model

The **device model**, shown in Figure 2.6, describes the overall structure of any physical control device that supports the execution of an IEC 61499 network of function blocks. The main purpose of the device is to provide an infrastructure to sustain one or more **resources**. An IEC 61499 resource has similar properties to the resource concept defined in the PLC programming languages standard IEC 61131-3 in that sense that a resource provides an execution environment for networks of function blocks. However, in contrast to IEC 61131-3, IEC 61499 resources are not necessarily directly bound to an execution unit (e.g. CPU). An IEC 61499 resource is a logical separation within a device and provides independent execution and control of networks of function blocks.

A device has a 'process interface' that provides the services that enable resources to exchange data with the Input and Output (I/O) points on the physical device. There is also a communications interface that provides communications services for resources to exchange data via external networks with resources in remote devices. As already depicted in Figure 2.5, a device may be interacting with more than one network. In such cases the communications interface handles the access to these as well. The internal structure and behaviour of the process and communications interfaces are not within the scope of IEC 61499 but they are expected to provide a range of services to support the execution of function blocks within the resources.

2.3.3 Resource model

The resource provides facilities and services to support the execution of networks of function blocks allocated to it. For this, the resource must ensure that events are used to schedule the appropriate functionality encapsulated within the function blocks at the correct priority and time. The resource is responsible for retaining the values of variables within function blocks between invocations. The resource is

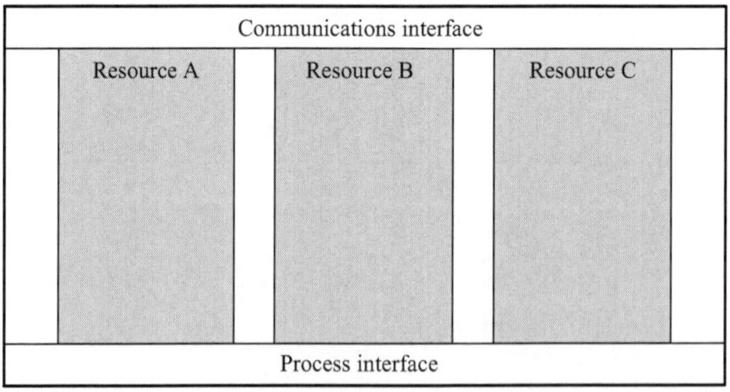

A device may have one or more resources, communication interfaces to other devices and a process interface.

Figure 2.6 Device model

also concerned with propagating events and transferring data between function blocks either on the same resource or between resources utilising communication services.

The resource provides interfaces to the communications systems and to the 'device specific services', i.e. to external services and sub-systems that are closely connected to the device, such as the device I/O sub-system or displays. For example, each resource will have an interface to the communications system to allow function blocks to exchange data with function blocks in remote resources and an interface to read and write to local device I/Os. The resource is therefore concerned with the mapping of data and event flows which pass between function blocks in the local resource to remote resource function blocks via the device communications interfaces. Similarly the resource maps all requests to read and write to device I/O onto the process interfaces.

It is clear that the resource defines the important boundary that exists between what is within the scope of the IEC 61499 model and what is device and system specific functionality. Issues such as operating system design and communications protocols that are specific to types of devices and networks are outside the scope of the standard.

Figure 2.7 depicts the main features of an IEC 61499 resource. Within the resource, it shows a network of interconnected function blocks linked by data and event flows. A scheduling function provided by the resource ensures that functionality encapsulated within function blocks is executed in the correct order, i.e. as required by the arrival of events at each function block. **Service interface** function

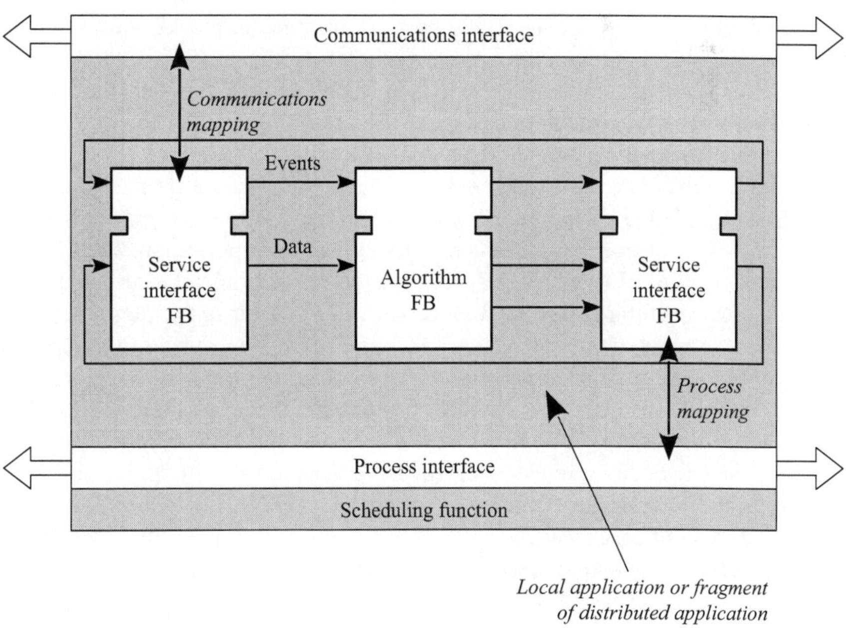

Figure 2.7 Resource model

blocks are a special form of function block that provide a link between function blocks and the interfaces of the resource. For example, a communications service interface function block can be used to read or send data to function blocks in remote resources. A number of different types of service interface blocks are identified and defined in IEC 61499 and are described in detail in Chapter 5.

An important characteristic of a resource is that it supports independent operation. A resource can be loaded, configured, started and stopped without affecting other resources in the same device or network. However, it is important to note that the management of a distributed application may require the co-ordinated control of a number of resources where function block network fragments are loaded, e.g. when loading and starting the various function block network fragments that make up the distributed application. The facilities required to achieve such co-ordination are an issue not fully addressed by IEC 61499.

2.4 Distribution model

The distribution model links the system independent application model with the system model. It is concerned with assigning application parts to devices and resources and to perform device specific configurations of the assigned application parts.

2.4.1 Mapping of applications

Applications can be 'distributed', that is, configured to run on several resources. A distributed application will consist of a network of function blocks with fragments of function blocks running on designated resources. Note that IEC 61499 considers distribution to concern the assignment of an application's function blocks onto different resources of the devices defined in the system model. For this assignment, the **mapping** statement is used. Figure 2.8 shows this mapping of our sample application from Figure 2.4 to the devices of the system model from Figure 2.5.

A specific feature of the IEC 61499 system and distribution model is that it allows devices to support the execution of more than one application. The standard defines a device model in which it is possible to load and unload distributed applications without disturbing existing applications; this is achieved through the use of management services within the device – these will be reviewed later in the chapter.

Because a device can support multiple resources, it is possible to separate the different applications within a device by mapping them to different resources (Figure 2.9). Furthermore it is also possible for an application to be mapped only to a single device, i.e. using one or several resources within the same device (e.g. Application 2 in Figure 2.9).

The timing and performance of an application will depend on the resources on which it has been distributed and the communications networks that connect them. For example, consider two identical applications for conveyor line control, running on two different networks of controllers, one network using fibre optic data links running at 1 GBs and the other using lower cost twisted-pair copper running at

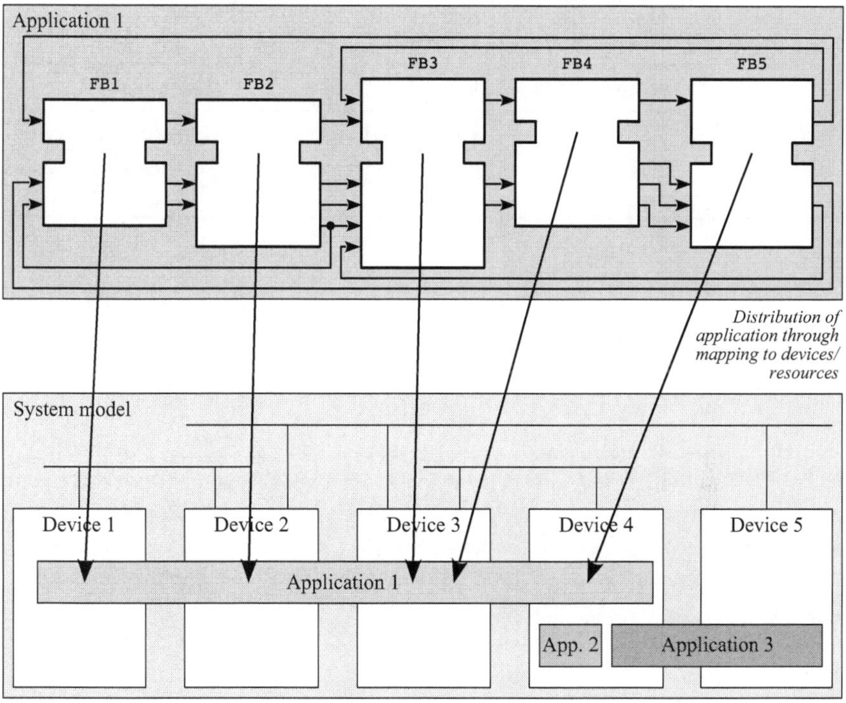

Figure 2.8 Distribution model

1 MBs. Because the communications data rates of the two networks are so different, the two applications will clearly have different performance and response times due to network latencies, although their internal software algorithms will be the same.

In contrast, function blocks are assumed to be 'atomic' and to only run in a single resource. In this respect, the performance of a function block is not affected by the characteristics of communications networks. However, the performance may to a

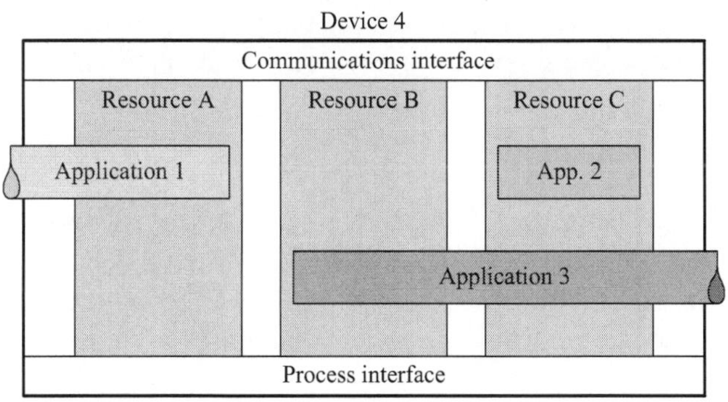

Figure 2.9 Application Resource Assignment

Table 2.1 Distribution characteristics

	Distributable	Timing	Reliability
Function block	No	Depends on device only	Depends on device only
Application	Yes	Depends on devices and communications	Depends on devices and communications

lesser extent be affected by the behaviour and characteristics of the resource and device in which the function block has been instantiated (e.g. computational power of the device). These distribution characteristics are summarised in Table 2.1.

2.4.2 Platform specific configurations

The mapping of the application results in two issues. The first is that distributing the application will break direct connections between function blocks (e.g. the connections between `FB4` and `FB5`). The second issue is that system specific configurations, such as the assignment of physical input port addresses to provide sensor values, are not part of the application model as the application model is platform and system independent.

In order to cope with this issue IEC 61499 allows the addition of system specific configuration values and extending the mapped application parts in the resources. The platform specific application extensions are typically the insertion of communication service interface function blocks and their configurations (e.g. communication channel identifiers). These communication service interface function blocks maintain the 'broken' connections across device and resource boundaries, as shown in Figure 2.10.

The need to perform these system configuration specific adaptations seems at first like an unnecessary overhead. However, it strongly supports the platform independence of the application and an application centered design process with the focus on the control functionality. The platform specific configuration late in the development cycle increases reusability of applications and application parts. Furthermore, it has the advantage that late changes in device and system configurations do not affect the application development to such a large extent as it is currently the case with the device-centric development approach of IEC 61131-3.

2.5 Management model

We have seen that a resource can support fragments of function block networks that form parts of distributed applications. But how are these application fragments created and managed? IEC 61499 was designed considering the requirements of adaptive reconfigurable automation systems. This requires that resources and their contained network of function blocks can be changed on the fly while the system is executing with no or only minimal disturbance to the controlled process. Therefore IEC 61499 defines a management interface with **management applications** and a

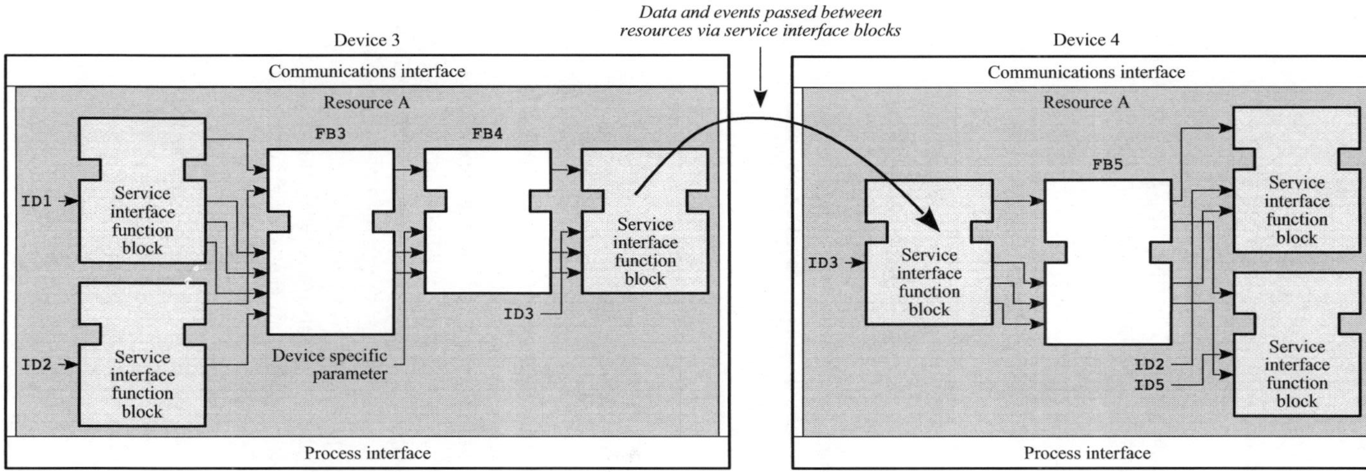

Figure 2.10 Device specific parameters and communication modelling

general behaviour model of the managed entities (i.e. resources, function block instances and their interconnections).

2.5.1 Management applications

IEC 61499 defines a special form of application called a management application that is responsible for managing the application life cycle of resources and of function block networks within a resource. The management application, with higher privileged functionality than normal applications, can construct/destruct parts of other applications by creating/deleting function blocks and connections. It typically will provide its management functionality via the communication interface to external agents such as remote programming stations.

The functionality of a management application needs to include:

- creating function block instances within a resource,
- creating data and event connections between function block instances,
- setting parameters of resources and function block instances,
- initiating the execution of function block instances as part of a distributed application,
- changing the operational state of resources and function block instances,
- providing services to support queries from communication links on
    - the status of function block instances including the operational state as well as the values of inputs and outputs,
    - the current function block instances and their connections,
- deleting function block instances,
- deleting data and event connections.

An important constraint is that the management application should be able to load function block network fragments for different applications without disturbing the execution of other running applications.

If the management application loads normal applications, how is the management application itself loaded? It is assumed that a device will need a certain level of baseline functionality within the management application in order to 'bootstrap' load the function blocks for the main applications. It is likely that parts of the management application will exist in a non-volatile form within the device and it is always possible to load applications when the device is powered-up.

Note: Some devices may hold all these function block networks in a non-volatile form that cannot be modified via external communications, in which case most of the functionality of the management application may not be required.

The standard proposes two schemes for providing management applications. Figure 2.11a depicts a device which has a special 'management resource' that contains management applications that provide functionality to build and maintain function block networks in the adjacent resources provided in the device. In Figure 2.11b an alternative arrangement is defined where each resource contains a management application that is responsible for loading function block networks within the same resource.

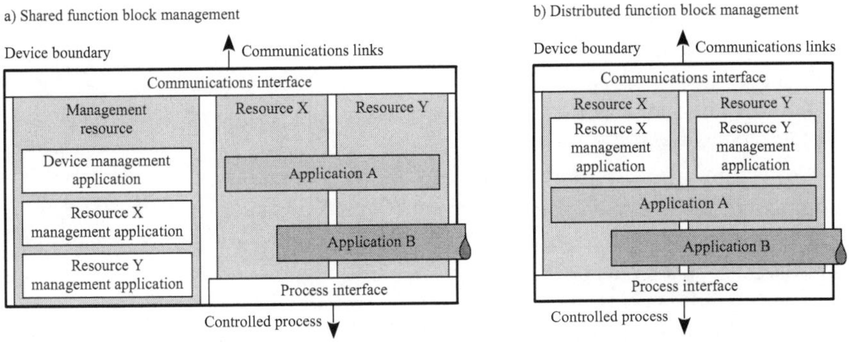

Figure 2.11 Management models

Management applications can be modelled in exactly the same way as other applications using networks of function blocks and service interface function blocks. It is likely that a management application will require a number of service interface function blocks to handle the interfacing with external communications links, e.g. to process requests to create function block instances. Furthermore, service interface function blocks will provide access to the execution environments for performing the management tasks. An example of such a function block and the required services is given in Section 5.5.

The standard also foresees situations where certain privileged functional units may take control of other function units. For example, a 'job loader' function block in device A may be used to load new function block definitions in remote devices B and C in order to modify an application for a certain type of batch profile. In such cases, there is a requirement to be able to give certain function blocks the rights to load, configure and change the states of other function blocks. The management applications that have been introduced in this chapter will clearly need to have special privileges to be able to load and start complete applications.

The full definition of the functionality of a management application is beyond the scope of IEC 61499. However, it clearly is important to load and configure IEC 61499 compliant devices in a consistent manner. The IEC 61499 compliance profiles as described in IEC 61499 Part 4 need to clearly define how the management functionality is implemented. The definitions on management applications as well as management function blocks (see Section 5.5) provide guidance on what is required. In Section 8.3 and Appendix B, more information on compliance profiles is given.

2.5.2 Operational state model of managed entities

Large systems based on networks of IEC 61499 function blocks will typically be developed, commissioned and brought on-stream in a series of stages. Furthermore, in such systems, changes, updates and maintenance occur frequently and should result in as little disturbance as possible. The execution model for function blocks as presented in Section 2.1.2 targets the normal operation state of applications and how events are handled during normal operation. However, during the other stages of a managed entity's life cycle, the handling of events or the correct processing of the data may not be possible. Reasons for this can be for example that the function block is not yet correctly parameterised, or input data is not yet valid because the data from the source is not available.

So, within a function block network, IEC 61499 proposes that the concept of an operational 'life cycle' is applied to functional units such as devices, resources and function blocks. This operational 'life cycle' is modelled in form of a common state machine which clearly defines for all managed entities when they are enabled for event handling, processing data and generating output data and events. The following common actions are performed in the different states:

IDLE: When entering the state IDLE, the managed entity is initialised. For example, the variables of a function block are set to their default values.

RUNNING: In this state, the managed entity is performing its normal operation. For resources, this means that events are delivered and function block execution is scheduled. For function blocks, this means that input events are processed according to the execution model presented in Section 2.1.2.

STOPPED: When entering this state, the currently triggered execution is finished, but after that no further execution is performed (e.g. a function block finishes its execution of actions triggered by an input event but will not accept any further input events).

KILLED: When entering this state, current execution is interrupted and no further execution is performed. This may leave the managed entity in an invalid state. Therefore, KILLED managed entities can only be deleted or reset to the initial state for further use.

Figure 2.12 shows the state machine with the allowed transitions between the states. The commands for transitioning between the states correspond to the

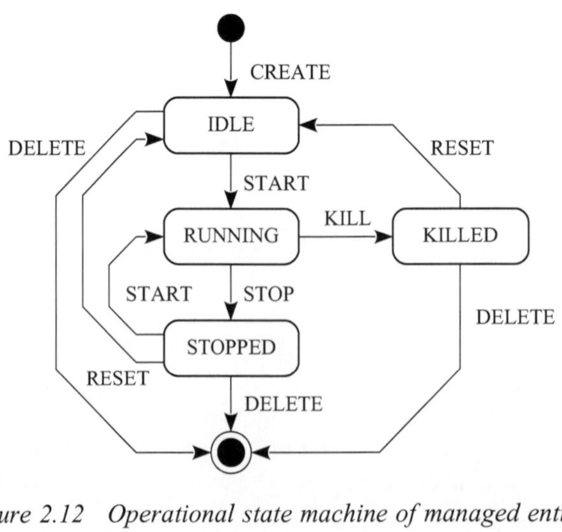

Figure 2.12 Operational state machine of managed entities

required functionality of management applications described in the last section. The management applications or the remote entities using them have comprehensive and well-defined execution control over all parts of an application. This allows not only a consistent start-up and shutdown of the automation system but also more advanced adaptation and reconfiguration scenarios as small application parts can be stopped, changed and restarted without influencing the other parts. How this infrastructure can be utilised to handle scenarios such as dynamic reconfiguration without the need to stop the overall automation systems is beyond the scope of this book but is explained in detail in [20].

To control and synchronise the loading of a distributed application, the standard proposes that strategies are developed to ensure 'a consistent operational state across its components'. For example, a distributed application may require that fragments of function block networks are loaded onto a number of different resources located in a variety of different and, in some cases, remote devices. The loading and configuration of the function blocks residing in the different devices may occur at different times. However, the switch from 'IDLE' to 'RUNNING' states for all associated devices and resources may need to be synchronised for an application to start-up correctly.

2.6 Exchange format for IEC 61499 entities

The ability to exchange all IEC 61499 models between different software tools and platforms is a very important aspect of the IEC 61499 standard. Being able to exchange function blocks between different products and systems will bring significant end-user benefits. This will include major cost savings from being able to reuse function block designs in different hardware and system configurations. IEC 61499 provides two options for supporting this: (i) a formal textual syntax and (ii) utilising the web-technology eXtensible Markup Language (XML).

2.6.1 Textual syntax for IEC 61499 entities

A significant feature of the IEC 61499 standard is a rich textual syntax that allows the models of entities such as applications, system configurations and function blocks to be described in a text language. The textual definitions can be used to unambiguously define models so that the graphical representations can be created automatically and consistently. For example, an application may be constructed from a network of several function blocks with appropriate parameters as well as data and event connections. All aspects of the design of the application can be represented using the IEC 61499 textual syntax.

It is envisioned that it will be possible to compile textual definitions as a means to check the validity of a particular model. The textual form is certainly useful for porting model designs from one workstation platform to another. It defines the semantic aspects of the model but does not fully describe the graphical details. For example, a particular model may have been developed showing function blocks located at certain positions in a network diagram. The actual location and finer

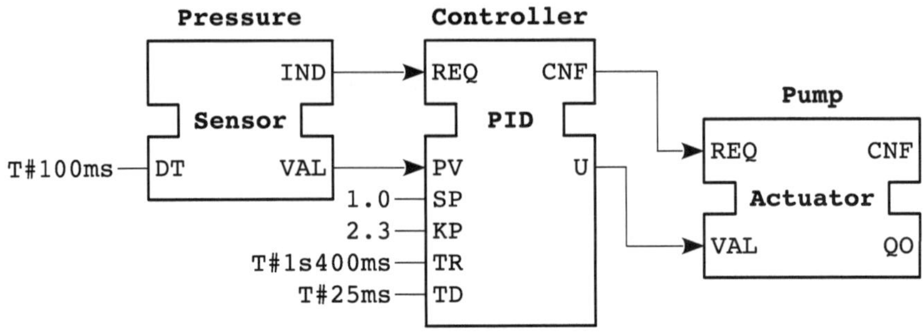

Figure 2.13 PID control loop application example

graphical details such as colour and font are not part of this syntax. However, the logical connections between function blocks are defined precisely. The standard contains an extensive annex that describes the production rules for all the features of the textual syntax.

Note: Some aspects of this syntax will be described in this book but for full and exact definitions it is advisable to refer, to Annex B in Part 1 of the IEC 61499 standard.

Application example

A simple application is used here to illustrate some of the features of the textual syntax. Many features of the syntax will be discussed in more detail in later chapters.

Consider an application which performs the closed loop control of the pressure in a vessel. A pressure sensor provides the current pressure value. The pressure in the vessel can be changed with a pump. This can be modelled with an application consisting of three function blocks as depicted in Figure 2.13. The `Sensor` function block provides cyclical updates of the current pressure value. This triggers the `PID` controller calculating the required value for the pump. The `Actuator` function block takes this value and applies it to the pump.

The textual syntax to describe the application example is as follows:

```
APPLICATION Controller (* Pressure control loop *)
  FBS
    Pressure : Sensor(
      DT := T#100ms);
    Controller : PID(
      SP := 1.0,
      KP := 2.3,
      TR := T#1s400ms,
      TD := T#25ms);
    Pump : Actuator;
  END_FBS
```

```
  EVENT_CONNECTIONS
    Pressure.IND TO Controller.REQ;
    Controller.CNF TO Pump.REQ;
  END_CONNECTIONS
  DATA_CONNECTIONS
    Controller.U TO Pump.VAL;
    Pressure.VAL TO Controller.PV;
  END_CONNECTIONS
END_APPLICATION
```

The following points can be noted in this example:

- Applications are identified by name, in our example this is `Controller`.
- Function block parameters are given in the function block instance declaration.
- Event and data connections are separated in own groups denoted by `EVENT_CONNECTIONS` and `DATA_CONNECTIONS`, respectively.

2.6.2 XML-exchange format

The ability to exchange software designs between different vendors' products has been a serious limitation of the IEC 61131-3 PLC software standard because IEC 61131-3 failed to include a full definition of a file exchange format. Especially, it has not been possible to exchange graphical PLC software designs using IEC 61131-3 languages between different PLC software support tools. In order to overcome this limitation, the PLCopen organisation developed an XML-based file exchange format [21]. Although this file format improved the situation, it still allows vendor specific extensions limiting the portability.

The shortcoming of having a purely textual syntax has been identified early in the development of IEC 61499. In order to overcome the limitations, XML has been considered as option long before PLCopen started their attempts. This effort resulted in the standardisation of XML structures in the form of Document Type Definitions (DTDs) in Part 2 of the IEC 61499 standard.

The XML format provides a means for storing not only the structural features of IEC 61499 models (e.g. function block instances and connections of an application), but also the graphical layout information (e.g. position of function blocks or bending points of connections) in a vendor neutral way. The portability of this file exchange format has already been demonstrated as IEC 61499 models have been exchanged between tools from different vendors. However, the XML format does not render the textual syntax obsolete as the textual syntax formally also defines the possible semantics of IEC 61499 models. This is currently not possible with the DTDs used for defining the IEC 61499 XML file structure.

Application example

The XML representation describing the pressure controller example from Figure 2.13 is as follows:

```
<Application Name="Controller" Comment="Pressure control loop">
  <SubAppNetwork>
    <FB Name="Pump" Type="Actuator" x="2358.8235" y="570.5882">
    </FB>
    <FB Name="Controller" Type="PID" x="1329.4117" y="382.35294">
      <Parameter Name="SP" Value="1.0"/>
      <Parameter Name="KP" Value="2.3"/>
      <Parameter Name="TR" Value="T#1s400ms"/>
      <Parameter Name="TD" Value="T#25ms"/>
    </FB>
    <FB Name="Pressure" Type="Sensor" x="476.47058" y="382.35294">
      <Parameter Name="DT" Value="T#100ms"/>
    </FB>
    <EventConnections>
      <Connection Source="Pressure.IND" Destination="Controller.REQ"
        dx1="247.05882" dx2="0.0" dy="0.0"/>
      <Connection Source="Controller.CNF" Destination="Pump.REQ"
        dx1="364.70587" dx2="0.0" dy="0.0"/>
    </EventConnections>
    <DataConnections>
      <Connection Source="Controller.U" Destination="Pump.VAL"
        dx1="164.70587" dx2="0.0" dy="0.0"/>
      <Connection Source="Pressure.VAL" Destination="Controller.PV"
        dx1="364.70587" dx2="0.0" dy="0.0"/>
    </DataConnections>
  </SubAppNetwork>
</Application>
```

The following points can be noted from this example:

- The overall structure of the XML representation is the same as the textual syntax.
- The textual syntax is more concise and more human-readable.
- The XML representation additionally contains the graphical layout information. In our example, this provides the position of function blocks (i.e. the attributes x and y) and the bend-points for connections (i.e. the attributes dx1, dx2 and y). This information has no impact on the functionality. In order to achieve independence between different drawing approaches (e.g. drawing resolution, font size) the base unit for coordinates is 1 per cent of the distance between two lines in the function block interface (i.e. the line height of the input or output ports).

Note: For better readability in this book we have used the textual syntax when explaining specific properties of IEC 61499 using a textual representation.

2.7 Summary

We have now reviewed the main concepts introduced by IEC 61499 that provide a framework and architecture to model function block oriented distributed systems. To summarise:

- The core element of IEC 61499 is the function block, which is executed in an event triggered manner. That means that function block execution is triggered by input events.
- There are both basic and composite function blocks to deal with different forms of function block construction and function block hierarchy.
- Service interface function blocks provide network communications and hardware interface facilities.
- Applications are developed independently from the control hardware by instantiating function blocks and interconnecting their event and data inputs and outputs.
- The system model defines a set of interconnected devices that can communicate via network connections.
- The device model supports one or more resources which provide support for the loading, configuration and execution of function block networks.
- An application can reside on one or more resources. Each resource can support the management and execution of part of an application; each part being distributed as a fragment of a function block network.
- With these models at hand, the IEC 61499 control system development work-flow consists of the following steps:

    1. Model the application by instantiating function blocks and interconnecting them.
    2. Model the system infrastructure by specifying devices, their resources as well as the communication links between the devices.
    3. Distribute the application to the modelled resources in the control devices.
    4. Perform platform specific parameterisation and insert communication function blocks.
    5. Deploy the distributed application to the devices utilising the management interface.

- The IEC 61499 provides two ways for exchanging the models in a portable way:

    1. The textual syntax provides a concise and formal specification method.
    2. The XML exchange format also includes the graphical layout information.

Chapter 3
Defining function block types

In this chapter, we will review how to create type definitions for function blocks and show how these can be used to create function block instances.

Specifically we will:

- *review different forms of function block definitions,*
- *show how event and data interfaces can be defined and how the external visible behaviour of function blocks can be described,*
- *examine how algorithms are constructed and linked to the event execution,*
- *consider how instances of function blocks behave,*
- *show how function block instances can be aggregated to composite function blocks.*

3.1 Types and instances

Before proceeding to describe the mechanisms provided in IEC 61499 for defining function block types in some detail, let us recall the role of function block types and instances. A function block type definition describes the external interface and internal behaviour of a particular type of function block. However, a function block instance is, in effect, a working copy of the function block created using the function block type definition. In a large system it is likely that function block type definitions will be held in various libraries for such purposes as 'control algorithms', 'alarm management', 'input sensors' and so on. Configuration of a large function block–based system will require selecting appropriate function block type libraries and then declaring function block instances based on selected function block types. Clear, precise and unambiguous definitions of function block types are therefore germane to using IEC 61499 effectively.

3.2 Different form of function blocks

The standard provides type definitions for two different kinds of function blocks, i.e. basic and composite; each kind has its own particular properties and uses – as listed in Table 3.1. The main characteristics of these two kinds are depicted in Figure 3.1. It should be noted that basic and composite function blocks always reside on a single resource and provide variables at

Table 3.1 Characteristics of the different forms of function blocks

Type	Definition	Comment
Basic Function Block	States defined using the Execution Control Chart (ECC). Algorithms defined using an appropriate language, e.g. Structured Text, Java, C.	Basic function blocks define the fundamental function blocks from which larger composite function blocks and applications can be built.
Composite Function Block	Constructed from a network of function blocks. It is defined by the data and event connections between component function blocks.	A composite function is built from a network of lower level function blocks. These can be either basic, lower level composite or service interface function blocks.

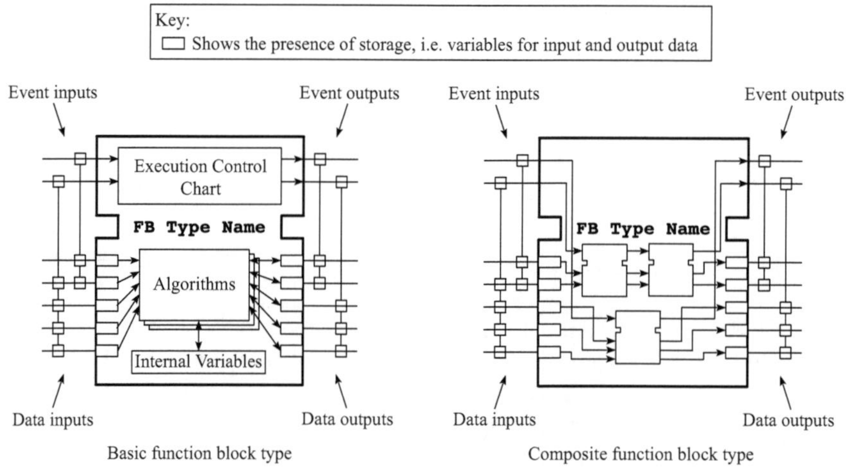

Figure 3.1 Different forms of function blocks

their inputs and outputs to hold data values. Basic function blocks also require internal storage for the execution control state machine. Composite function blocks in contrast require storage for the internal function blocks and connections.

3.3 External interface declaration

All function blocks have in common the external interface which defines how to interact with it in applications and composite function blocks. IEC 61499 separates the static interface structure (i.e. the event and data ports) and the definition of the dynamic interface behaviour (i.e. how to use the interface).

3.3.1 Static interface declaration

The static interface declaration as shown in Figure 3.2 has the following features:

- The function block type name should be positioned in the centre of the main block as shown by `FB_TYPE_NAME` in Figure 3.2.
- Inputs to the function block are always shown on the left of the function block; outputs are always shown on the right side of the function block.
- Input events are depicted entering the left side, output events are shown coming out from the right of the upper part of the function block.
- The names of input and output variables are shown inside the function block next to their associated graphical connectors.
- The data types of input and outputs are shown at the left-hand and right-hand ends of the graphical connectors, respectively.

The graphical representation provides sufficient information to be used as a formal type declaration. In fact, a primary objective of IEC 61499 is that graphical representations always have a precise textual representation (textual syntax as well as XML). It is envisioned that graphical modelling tools will always be able to convert graphical forms into textual representations and vice versa.

Input events such as the `EI1` event depicted in Figure 3.2 that are shown entering the function block header can, if required, be associated with one or more input variables. The association of input events with input data uses a construct that IEC 61499 denotes as the `WITH` qualifier. In the graphical representation this is shown using small square connectors that link the event with its associated data.

This would typically be required where the function block needs to sample input values prior to the execution of the function block. In the function block example, inputs `DI1` and `DI2` need to be sampled whenever an `EI1` event occurs. Similarly, the `DI2`, `DI3` and `DI4` input values will be sampled when the `EI2` event occurs.

It is also possible to associate output events with certain output variables. This is used to denote those output variables that have been updated by an internal algorithm and are ready at the instant the output event is fired. In the function example in Figure 3.2 the `EO2` output event occurs when the function block has

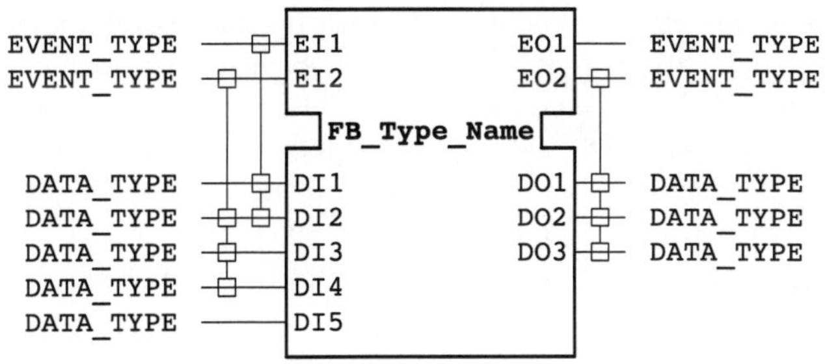

Figure 3.2 Graphical interface declaration for function block types

updated outputs `DO1`, `DO2` and `DO3`. The same `WITH` textual qualifier and graphical representation are used to show the association between output events and their output data.

Note that every input and output of a function block should be associated with at least one event input or output, respectively. This is because at least one event is always required to signal when an input value is sampled or when an output value has changed. Another way of viewing this is to regard the event and its associated data as a type of message that allows an event and its data to be passed between function blocks as a coherent set. If a data input or output is not associated to any event (e.g. `DI5` in Figure 3.2), it cannot be used for a connection. It can be seen as constant value that gets a fixed parameter value which is not changed during the execution of the application.

One consequence of this feature is the implication that a function block must have storage to hold the values of inputs between event samples. Likewise, it has storage to hold values of outputs between times when output events are fired. Of course, there is always the possibility that a function block may receive events with data at a rate that is faster than the function block can store and then process. The standard states that, in such cases, the underlying scheduling function should prioritise function block execution in a manner that ensures that such overload situations do not occur, as described in Chapter 2[1].

Events can be defined to have an optional event type. This is provided so that function blocks can be designed to only accept certain types of events at their event inputs. Event types provide increased robustness in the design. This is a logical extension to the way that data types are used to increase design integrity by only allowing the connection of input and outputs carrying compatible data. Similarly event types ensure that only events of particular type can be used for a compatible purpose.

If a function block's event input or output is not given an event type, the default type `EVENT` is applied. Events of this generic type `EVENT` can be connected to any other event inputs of type `EVENT`. Conversely, an event input of type `EVENT` can receive events of any event type.

3.3.2 Defining the dynamic interface behaviour

With the static interface as described in the last section, a function block is in principle fully defined to be usable in any application and composite function block. However, the important point that is missing is a description on how to use the interface (e.g. in which order to send input events, what input data it expects with these events and what outputs will be provided by the function block). In order to allow the description of the dynamic behaviour, IEC 61499 defines the use of service sequence diagrams with every function block type declaration. Service sequence diagrams are based on time-sequence diagrams from the ISO Technical Report 8509. These can be used to show the timing and sequential relationships

[1] [20] has a detailed description of how to provide deterministic real-time constrained execution of IEC 61499 function blocks and function block networks.

between various interactions with the function block. In fact, time-sequence diagrams are commonly used in communications standards and provide an effective way to visualise the order in which various messages or events occur. In software modelling sequence diagrams are used to model the interaction of objects (e.g. UML [23]).

For a function block definition, zero or more service sequence diagrams can be used to describe different usage and execution scenarios for the function block. Service sequence diagrams have only a descriptive task in IEC 61499, therefore they may be omitted. An IEC 61499 service sequence diagram is structured as depicted in Figure 3.3. Each service sequence diagram is identified with a name shown in the top middle of the diagram. It consists of two interfaces separated by two vertical lines. Typically the left interface is denoted to the FB interface and the right to the FB internals. A service sequence consists of one or more service transactions which are ordered from top to bottom. Service transactions consist of one or more service primitives. A service primitive is either an input primitive (i.e. event going into the interface) or an output primitive (i.e. coming out of the interface).

Typically, service sequences describe the execution order in a strict order one after each other. Sometimes the exact order of the service transactions is not needed. Then, it is only necessary that a set of service transactions occurs in any order before the next transaction is valid. This can be denoted by inserting a '∼' between the service transactions. For example, in the service sequence `Rendezvous` in Figure 3.4 the events `EI1` and `EI2` have to occur once before the output event `EO2` is sent. Typically events alone do not describe a function block's behaviour. The associated data elements and their values also define it. Therefore, data associated with events as well as their values can be defined in service sequences (e.g. the `DataSequence` in Figure 3.4) as well.

Further details on the function block behaviour can be provided by adding service primitives on the right side – the function block internals – to service sequence diagrams. For example, the `InternalDescription` service sequence in Figure 3.4 implies that the input event `EI1` will trigger the initialisation of the function block and when the initialisation is finished the output event `EO1` is raised. Service transactions with only one service primitive can be used to describe a series

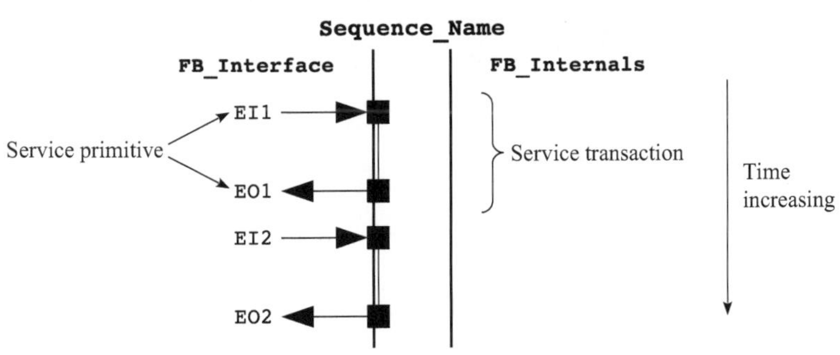

Figure 3.3 Overview service sequence diagram

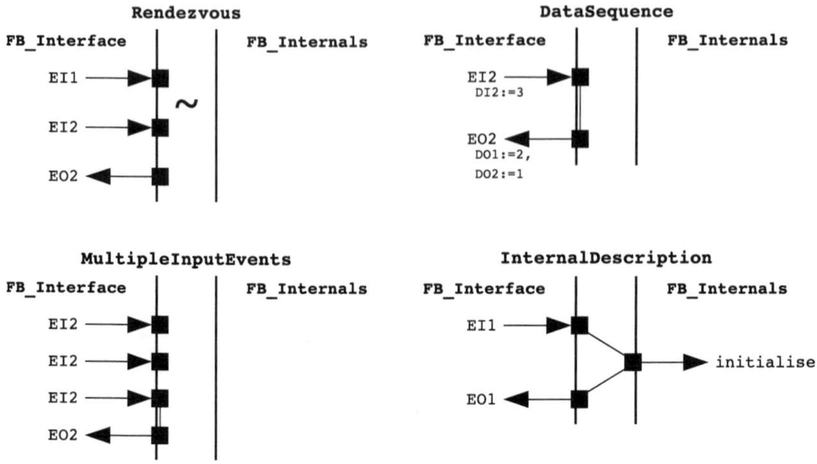

Figure 3.4 Service sequence diagram details

of input or output events that have to occur in sequence before the next step is valid. For example, in the `MultipleInputEvents` service sequence shown in Figure 3.4, `EI2` has to occur three times before `EO2` is issued once.

It should be noted that currently service sequences only have a descriptive task providing documentation for the users of the function block. However, it is envisaged that service sequences can also be used as the basis for formal analysis (e.g. is the function block correctly implemented or is it correctly used).

3.3.3 2 out of 3 voter example

A simple function block is used here to illustrate the external interface declaration for function blocks. Consider a function block that applies a '2 out of 3 voting' algorithm on three inputs `A`, `B` and `C`. The static interface of this function block can be modelled as depicted in Figure 3.5. The function block has two event inputs, `Vote` and `Reset`.

The `Vote` event is used to trigger the voting process which checks the state of the three Boolean inputs `A`, `B` and `C`. If two or more inputs are `TRUE`, the output `State` is set to `TRUE` and remains `TRUE` until `Reset` is triggered. When the voting is completed, an event is produced at the `Voted` event output. The `Reset`

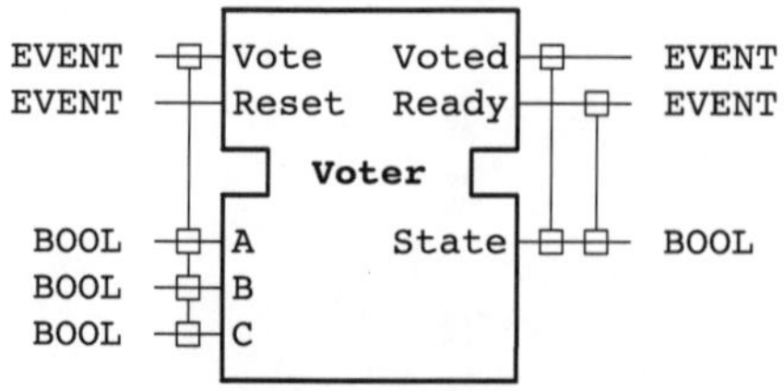

Figure 3.5 Interface of the Voter function block example

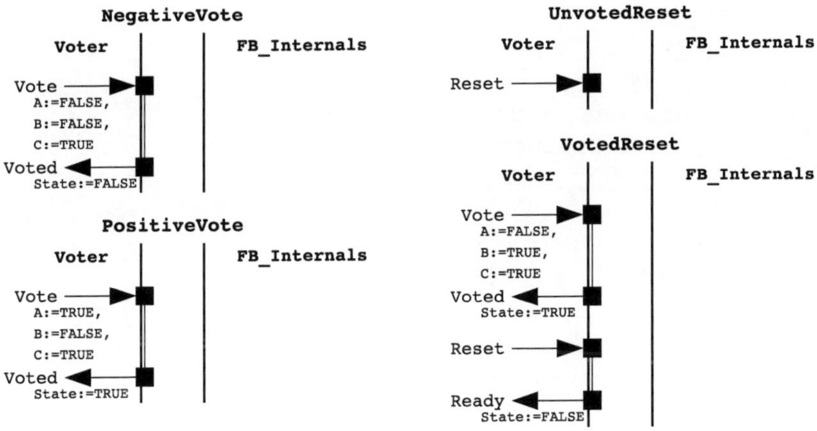

Figure 3.6 Service sequence diagrams describing the Voter function block's behaviour

event triggers a reset of the `State` output and after that an event is produced at the `Ready` output.

Service sequence diagrams describing this behaviour in a more formal way are shown in Figure 3.6. The service sequence diagrams describe four concrete scenarios. However, these do not cover all possible scenarios. In order to capture these, a specification of all different possible input combinations and succeeding event flows would also be required. This would result in quite a number of service sequences. However, the subset presented here is sufficient to describe the behaviour for a potential user.

The textual syntax to describe the interface of the `Voter` function block is as follows:

```
FUNCTION_BLOCK VOTER (* Voter FB *)
  (* Event definitions *)
  EVENT_INPUT
    Vote WITH A,B,C;  (* Trigger Voter *)
    Reset;            (* Reset event *)
  END_EVENT
  EVENT_OUTPUT
    Voted WITH State;
    Ready WITH State;
  END_EVENT
  (* Variable definitions *)
  VAR_INPUT
    A : BOOL;
    B : BOOL;
    C : BOOL;
  END_VAR
```

```
  VAR_OUTPUT
    State : BOOL;
  END_VAR
  (* Service sequence definitions *)
  SERVICE Voter/FB_Internals
    SEQUENCE NegativeVote
      Voter.Vote(A:=FALSE, B:=FALSE, C:=TRUE) ->
          Voter.Vote(State:=FALSE);
    END_SEQUENCE
    SEQUENCE PositiveVote
      Voter.Vote(A:=TRUE, B:=FALSE, C:=TRUE) ->
          Voter.Vote(State:=TRUE);
    END_SEQUENCE
    SEQUENCE UnvotedReset
      Voter.Reset();
    END_SEQUENCE
    SEQUENCE VotedReset
      Voter.Vote(A:=False, B:=TRUE, C:=TRUE) ->
          Voter.Vote(State:=TRUE);
      Voter.Reset() -> Voter.Reseted(State:=FALSE);
    END_SEQUENCE
  END_SERVICE
END_FUNCTION_BLOCK
```

The following points can be noted from this example:

- The textual definition includes `EVENT_` and `VAR_` sections to declare all the events and data for the input and output interfaces of the function block.
- The assignment of data points to events is performed at the event declaration using the `WITH` statement followed by the data point names.
- The keywords `SERVICE` and `END_SERVICE` are provided to contain definitions of the service sequence. As part of the `SERVICE` statement, the left and right interface names are defined. The left and the right interface names are separated by a '/' symbol.
- Each service sequence is introduced with the keyword `SEQUENCE` and closed with the key word `END_SEQUENCE`. The sequence definition defines the service primitive and associated parameters that initiate the transaction.
- The '->' operator implies two things: (i) that there may be a time delay due to external processing or communications latency and (ii) that the resulting service primitive, as stated to the right of this operator, will occur as a consequence of the service primitive on the left.
- The names of sequences must map one-to-one to the names given on the service sequence diagrams. The names of inputs which should be valid when the service is initiated are enclosed in parentheses after the service primitive. Similarly, the names of outputs which are set for the resulting service primitive are shown on the right.

3.4 Defining basic function blocks

There are four main elements that together depict the properties and behaviour of a basic function block: (i) the external interface declaration as explained in the last section, (ii) internal variables, (iii) algorithms and (iv) the Execution Control Chart (ECC) that defines the relationships between events, states and algorithm execution as depicted in Figure 3.1.

3.4.1 Internal behaviour

There are two aspects to the description of the internal behaviour of a basic function block – algorithm bodies and function block execution control. The basic function block will generally contain one or more algorithms. Each algorithm is invoked by the resource scheduling function in response to particular input events arriving at the function block interface. As the algorithm executes it is allowed to process data from input, output and internal variables to create new values for both internal and output variables. Internal variables can be used to capture algorithm results that are not applicable as outputs at the function block interface. Internal variables like outputs are always retained between function block invocations (e.g. integral or derivative parts of a PID algorithm). When an algorithm has completed its execution it may fire output events to signal that output data is ready and can be 'consumed' by other function blocks.

Note: Algorithms cannot access any data outside of the function block, which is an important property to ensure that basic function blocks are self-contained and behave the same way independently from the context where they are used.

IEC 61499 does not define a specific language that should be used for algorithm definitions. Any high-level language can be used where a mapping between the input and output variables and their data types and variables within the algorithm's language can be defined and provided. We will see later that structured text (ST), as defined in the PLC Languages Standard IEC 61131-3, as well as Java are both good examples of high-level languages that can be used to express the behaviour of function block algorithms.

An important aspect of the behaviour of a basic function block concerns modelling the relationship between events and algorithm execution. This is achieved using a concept called the **Execution Control Chart** (ECC). Like other features of the function block, the ECC can either be defined graphically or textually.

Each basic function block requires an ECC to define the following:

- the internal states of the function block,
- how the function block will respond to each input event,
- which algorithms are activated in which order in response to input events,
- which output events are fired when associated algorithms have finished their execution in response to specific input events and the ECC's current state.

The ECC is a form of state transition diagram that bears several similarities to the graphical Sequential Function Chart (SFC)[2] in IEC 61131-3. However, as a

[2] SFC – Sequential Function Chart language defined in the PLC Programming Language standard, IEC 61131-3 and used to program sequential behaviour in terms of steps and transitions.

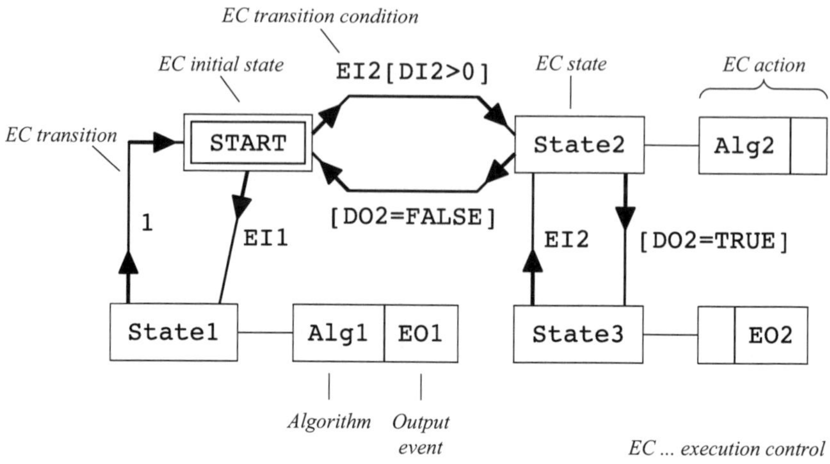

Figure 3.7 Execution control chart

state modelling technique its purpose is very different in behaviour and function and should not be confused with graphical programming languages such as SFC.

Figure 3.7 is an example of an ECC showing its main features. The interface elements are taken from the generic function block interface presented in Figure 3.2. In this case, the ECC depicts four states `START`, `State1`, `State2` and `State3` that correspond to the four states of the function block. The `START` state represents the initial state of the function block when it is created. When entering `State1` the algorithm `Alg1` will be executed and the output event `EO1` is issued. `Alg2` is executed when entering `State2` and when `State3` is entered the output event `EO2` is issued.

The transition between states can be depicted by either a) an event, b) an event with a Boolean expression, c) a Boolean expression and no event or d) no event and be unconditional (i.e. always true). Table 3.2 lists the transition conditions for this example ECC.

In this example, `State1` and `State2` are transient states that are only active while their associated algorithm (i.e. declared in the EC action) is executing.

Table 3.2 Transitions in the example execution control chart

From State	To State	Transition condition	Transition description
START	State1	EI1 ()	Event
State1	START	1	Always true
START	State2	EI[DI2>0]	Event and Boolean expression
State2	State3	[DO2 = TRUE]	Boolean expression
State2	START	[DO2 = FALSE]	Boolean expression
State3	State2	EI2	Event

In more complex function blocks it is possible for states to represent fundamental states or modes of the function block.

3.4.2 Execution control chart features

The standard defines a range of features and rules that apply to using ECCs which are summarised as follows:

- A basic function block will always have exactly one ECC, which can be defined graphically or using the textual syntax in the **execution control block** section of the function block type definition.
- Each ECC must have an initial state that is always depicted with a double-outlined shape.
- Each ECC has one or more states where either round or rectangular single-outlined shapes can be used to depict states within the ECC.

  Note: Throughout this book, rectangular shapes are used to represent ECC states.
- Each ECC state can have zero or more actions. Each action is normally associated with one algorithm and one output event. When entering a state, all actions defined for the state are executed one after each other. However, an action may have a null algorithm where it is only required to fire an output event. It is also possible to have actions with no output events. Typically a state will have at least one action that has an output event to signal to the external world that certain outputs have been updated.

  As an example, Figure 3.8 depicts a state `State` that has three actions to call algorithms `Alg1`, `Alg2` and `Alg3`; the first and the last trigger output events `EO1` and `EO2`, respectively when they complete execution.
- The ECC can utilise but not modify variables declared within the function block (i.e. input, output and internal variables).
- Transitions define the possible state changes. Transitions are guarded by a transition condition consisting of an input event, a guard condition, or both. A guard condition is a Boolean expression utilising input, output and internal variables.
- Several transitions can have the same state as target. Furthermore, several outgoing transitions with different transition conditions can be associated with one state. Outgoing transitions are prioritised by their order of creation or order of definition in the textual syntax. Tools may show this order in form of numbers at the transition origin as shown in Figure 3.8.

It is important to note that transition conditions within an ECC are not tested until the state preceding the transition is active. So, even though a function block may have a very complex ECC with a large number of transitions between states, the overhead involved in testing transitions may be small because at any time only one state will be active.

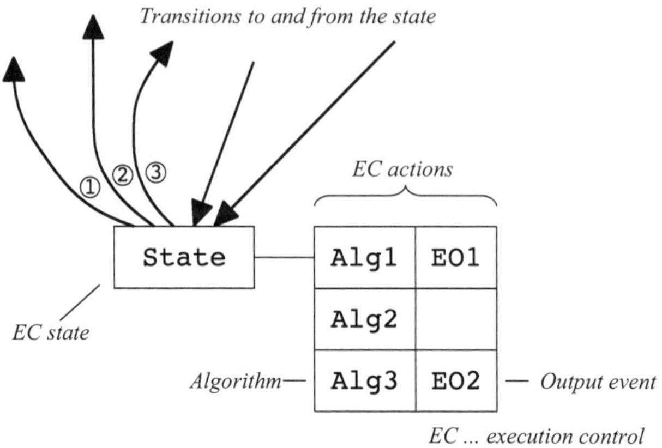

Figure 3.8 Execution control state example

It should be stressed that the ECC is primarily intended to represent the relationships between input events, algorithm execution and firing of output events. Originally the intention of IEC 61499 was to not use the ECC for modelling application state behaviour, e.g. to model various control modes. However, using the ECC for modelling application state has the great advantage that it is an efficient way for representing it in an easy understandable form. Furthermore it greatly reduces the complexity of algorithms. Therefore, the ECC has become more important for modelling the behaviour of basic function blocks.

3.4.3 Execution of basic function blocks

Note: In the 2nd edition of IEC 61499-1 the definitions for executing basic function blocks have been refined and further detailed in order to remove several semantic ambiguities. The following description is based on this new, refined execution model.

The standard defines very precisely the manner in which events that arrive at basic function block event inputs trigger changes of state within the ECC that then, in turn, cause the scheduling of function block algorithms by the resource and finally the generation of output events. The main aspects of this behaviour are summarised as follows:

- The resource maintains event occurrences in applications and delivers input events to the function block instance. Here, the resource has to ensure that no more than one event is delivered at any time.
- The resource treats function block execution as atomic. That means a new input event is only delivered after the execution triggered by the previous input event has finished.
- The resource then only allows a change in state within the ECC if the transition condition leading from the current active state (i.e. from the transition's predecessor state) is fulfilled.

The execution behaviour of basic function blocks for the evaluation and execution of the ECC can be described with a simple state machine consisting of three states as shown in Figure 3.9. The four transitions (t1 to t4) are taken (i.e. leading to a state change) on the following conditions:

t1: An input event is delivered by the resource. Any input variables associated with the event using the `WITH` construct are sampled and stored within the function block.

t2: All outgoing transitions of the currently active state are evaluated to `FALSE` and therefore no EC transition is crossed.

t3: A transition condition evaluated to `TRUE`, resulting in the crossing of an EC transition and leading to a new active EC state.

t4: All actions of the activated state have finished their execution.

The operations performed when the three states of the ECC operational state machine are activated are described in Table 3.3. Here, the main execution is performed in the states s1 and s2. State s1 is activated either when an input event is delivered by the resource (t1) or when the execution of EC actions has finished (t4). On the activation of the state s1, the EC transition conditions of the currently active EC state's outgoing EC transitions are evaluated. The first EC transition for which the EC transition condition evaluates to `TRUE`, is crossed. The following rules apply:

- An EC transition is crossed by the deactivation of its predecessor EC state and the activation of its successor EC state.
- The order in which EC transition conditions are evaluated corresponds to the order in the textual syntax, the XML representation or the order shown in the graphical representation (see Section 3.4.2).
- The input event occurrence which activated the function block execution is only considered in the evaluation of EC transition conditions if state s1 was entered via t1. That means that:
    - On the entering via t1, only EC transition conditions using the current input event or transition conditions without an input event will be evaluated.

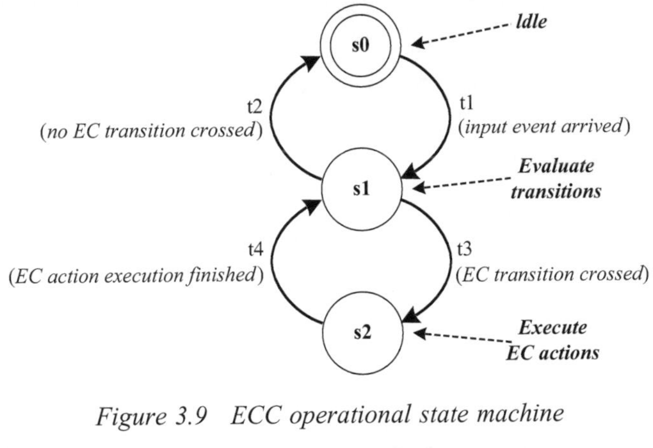

Figure 3.9 ECC operational state machine

- On entering state s1 via t4 (i.e. after the execution of EC actions has been completed) only transition conditions with no input event will be evaluated.

On entering state s2 the EC actions, if any, of the newly activated EC state are executed. This is performed sequentially, one EC action after each other. For each action, first the algorithm is executed and after its completion the associated output event is issued. If an action does not contain an algorithm or output event, the respective element is ignored. For example, if no algorithm is specified the event is directly issued. The order of action execution is according to the graphical representation from top to bottom, which is equivalent to the order in which they are declared in the textual syntax or in XML.

Note: Because t4 goes back to state s1, the EC transitions are evaluated recursively after an activated EC state's EC actions are completed. The function block execution is ended eventually when no transition is crossed any more (i.e. transition t2 back to state s0).

The way in which a resource actually executes a particular algorithm of a basic function block is to some degree 'implementation-dependent'. As stated earlier, an algorithm can be expressed using a variety of languages and therefore its behaviour is not regarded as being within the scope of IEC 61499. The algorithm implementation should, however, have the following characteristics:

- Input, output and internal variables should be mapped precisely and unambiguously to variables within the algorithm.
- The algorithm should be encapsulated in a way that it can only read and write to variables within the function block body so that the encapsulation property of function blocks is fulfilled.
- The algorithm execution time should be short in relation to the expected arrival rate of events that will trigger its execution. Clearly, if an algorithm is designed to execute every 100 ms and is going to be triggered using a 100 ms clock event, its worst case execution duration should be well under 100 ms to allow other algorithms to execute.
- The algorithm should have a well-defined initial state, which it can enter when the function block is first made ready for execution by the resource.

Table 3.3 Primary states of basic function block execution

State	Operation	Notes
s0	Idle	In this state, nothing in the function block is either executing or pending execution. The function block is waiting for the arrival of input events.
s1	Evaluate transitions	The transition conditions of the currently active EC state's outgoing transitions are evaluated. The first transition condition which evaluates to `TRUE` is crossed, which means the currently active EC state is deactivated and the target state of the transition is activated.
s2	Execute actions	The EC actions of the activated EC state are performed.

One consequence of this IEC 61499 execution model for basic function blocks is that no input events are delivered during the execution of a basic function block. A new input event (the same or another) is only delivered after the basic function block has finished its execution. Resources need to provide means to correctly deliver events occurring during function block execution (e.g. as described in [20]). However, one should note that if more events are triggered, than can be handled by the function block, an overload situation will result. For this reason, the standard states that the resource should provide means to detect when such overload conditions exist and take appropriate action for error recovery.

Note that it should always be possible, by various means, to create a model where event overloads are avoided; for example, by creating function blocks to provide event and data queues, or by feeding back loading information to upstream function blocks to modify event output generation rates.

3.4.4 Behaviour of instances of basic function blocks

Using either textual or graphical declarations, it is possible to define instances of basic function blocks, created from basic function block type definitions in applications or composite function blocks. An instance of a basic function type will have behaviour that is defined by its type definition but it will have its own storage for its input, output and internal variables and for holding the state of its ECC. In other words, the state of each basic function instance is totally independent of the other instances.

The resource in which a basic function block has been declared will initialise each basic function block instance prior to activating the function block for the first time. This initialisation includes:

- Setting the value of each input, output and internal variable to an initial value as defined in its declaration given in the function block type definition. Where an initialisation value has not been given, default values for the particular data type will be taken. For example, Boolean inputs that do not have a defined initial value will always be initialised to `FALSE` – the default initial value for Boolean (`BOOL`) variables.
- Algorithm internal states will be initialised. For example, if an algorithm has been defined using the IEC 61131-3 Sequential Function Chart (SFC) language, then the algorithm will be reset so that it will start on the SFC initial step.
- The initial state of the instance ECC will be set active; all other states within the ECC will be inactive.

3.4.5 Basic function block examples

Two out of three voter implementation

The first example, the `Voter` function block, as already used in the interface specification section (see Figure 3.5), a will be implemented as a basic function block. In order to model the described behaviour, the states and transitions depicted in Figure 3.10 are used. Upon a `Vote` input event in the start state `Ready`, the `Vote`

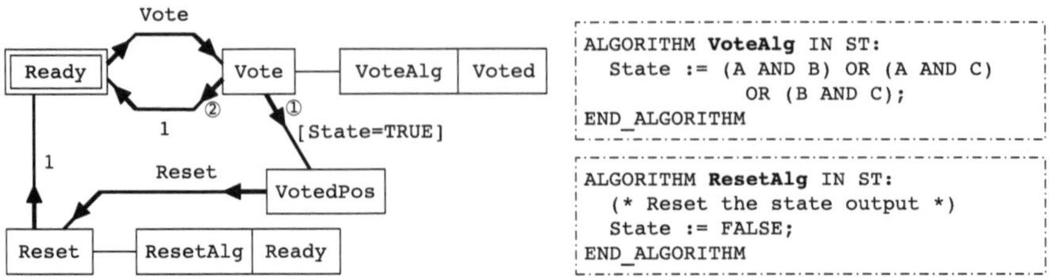

Figure 3.10 Voter example execution control chart

state is entered and the algorithm `VoteAlg` is executed. This algorithm performs the voting and after its completion it triggers the `Voted` output event to inform succeeding function blocks about the voting result. If the vote resulted in a positive result, the state `VotedPos` is activated; otherwise we go back to the `Ready` state. When the `VotedPos` state is active, a `Reset` input event will lead to a transition to the `Reset` state – invoking the `ResetAlg`, triggering the `Ready` output event and finally going back to the `Ready` state waiting for the next vote request. The algorithms are expressed using the IEC 61131-3 Structured Text language as shown in the right part of Figure 3.10.

The textual syntax to describe the `Voter` function block's internal structure is as follows:

```
FUNCTION_BLOCK VOTER (* Voter FB *)
  EVENT_INPUT
    (* Interface definition omitted, can be found in last
       section *)
  END_VAR

  EC_STATES
    Ready;
    Vote : VoteAlg -> Voted;
    VotedPos;
    Reset : ResetAlg -> Ready;
  END_STATES

  EC_TRANSITIONS
    Ready TO Vote := Vote;
    Vote TO VotedPos := [State = TRUE];
    Vote TO Ready := 1;
    VotedPos TO Reset := Reset;
    Reset TO Ready := 1;
  END_TRANSITIONS

  ALGORITHM VoteAlg IN ST;     (* Voter algorithm *)
    State := (A AND B) OR (A AND C) OR (B AND C);
  END_ALGORITHM
```

```
  ALGORITHM ResetAlg IN ST;    (* Reset Algorithm *)
    State := FALSE;    (* Reset the state output *)
  END_ALGORITHM
END_FUNCTION_BLOCK
```

The following points can be noted in this textual syntax example:

- The internal states of the function block are defined and associated with the events that trigger transitions between states; this is declared in the sections `EC_STATES` and `EC_TRANSITIONS`.
- Actions consisting of algorithms and output events are defined in the `EC_STATES` section of the function block definition.
- Each algorithm is defined in an `ALGORITHM` section, where the type of language is specified and followed by the body of the algorithm expressed using the language. In this example both the `ResetAlg` and `VoteAlg` algorithms are defined using the Structured Text (ST) language. However, note that IEC 61499 does not preclude the use of other languages such as JAVA or C to define the algorithm contents. In fact, surprisingly, IEC 61499 does not define the textual syntax for algorithm definition but allows any existing standard textual language to be used.

Ramp generation

In the first basic function block example the ECC was the main element for capturing the function block state. In contrast, in the second example, we consider an algorithmic basic function block where the functionality is mainly defined in the algorithms. Let us consider the required behaviour for a very simple `Ramp` function block. This function block ramps an output `Out` from an initial value given by input `X0` to a target value `X1` during a time given by the `Duration` input. The `Cycle` input defines the elapsed time between updates of the Ramp output. The function block also checks whether the output exceeds the input `PV`, in which case the `Hold` output is set to `TRUE`. It is assumed that the `Ramp` function block is called repeatedly and at an update rate given by the `Cycle` time, for example, it may be configured to execute every 200 ms.

The graphical type definition of this basic function block consisting of the static external interface declaration, the internal variables, the execution control chart and the algorithms is given in Figure 3.11. The behaviour of the `Ramp` function block is now completely defined with this type declaration.

On the arrival of the initialisation event `INIT`, the input values that characterise the `Ramp` behaviour, i.e. `X0`, `X1`, `Cycle` and `Duration`, are stored. The `INIT` state is activated causing the initialisation algorithm `InitAlg` to be invoked. This resets the internal timer variable `T`. The output event `INITO` is fired when the initialisation algorithm has terminated.

Similarly, when the run event `REQ` is received, a transition to state `Ramp` occurs causing the `RampAlg` algorithm to run. This calculates the new output value for `Out` based on the values of `X0`, `X1`, `Cycle` and `Duration` and the time into

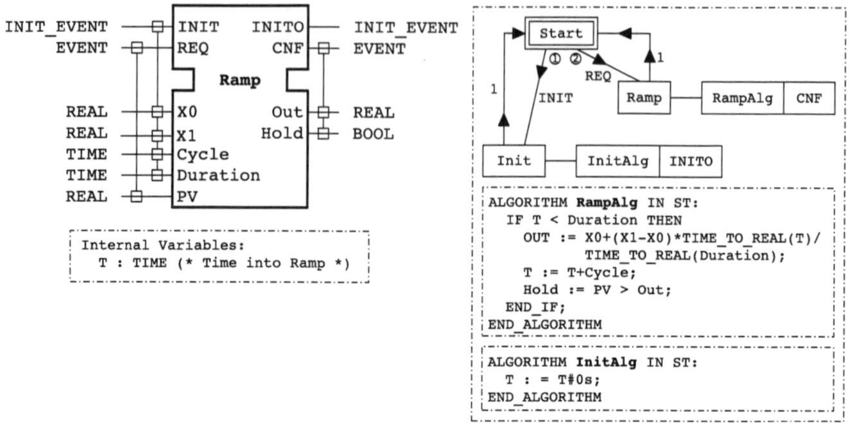

Figure 3.11 Ramp basic function block example

the Ramp T. The algorithm also checks whether the output exceeds the input value of PV, in which case the output Hold is set to TRUE.

The textual syntax for this function block is left as an exercise to the reader, who may refer to the textual syntax from the last example and the information given in Appendix E.

3.5 Defining composite function blocks

Composite function blocks provide a means to build more complex function blocks from basic, other smaller composite and service interface function blocks in a hierarchical fashion. The type definition for a composite function block contains declarations of function block instances of selected types that are linked by data and event connections. The standard regards function blocks that are used within a composite function block as **component** function blocks. The data connections between component function blocks define the transfer of data values between function block outputs to inputs, while the event connections define the order of execution of function blocks within the composite function block.

3.5.1 Rules for composite function block type specification

There are a number of rules and restrictions regarding the specification of composite function block type definitions, particularly regarding the connection of the external event and data inputs and outputs with the internal component blocks. These rules arise because events cannot be fanned-out directly; i.e. there must be a one-to-one connection between an event output and an event input. It is possible to make an event generate multiple concurrent events by using an E_SPLIT function block; this is one of IEC 61499's special event function blocks discussed in Chapter 6. In contrast, data inputs can be fanned-out allowing a single data output to drive many different data inputs.

Rules for event connections

The following rules apply to the connections between composite event inputs and outputs and component inputs and outputs.

- Each composite function block's event input must be connected to exactly one input event of an internal component function block or must be 'through routed' to a composite function block event output. That is, it is not possible for the event input to be connected directly to multiple event inputs in different component function blocks.

  Note: Event function blocks such as `E_SPLIT` or its shorthand notation can be used to create fan-out events from a single input event if required, see Chapter 6.
- Each component function block's event input must be connected to exactly one component function block output event or to a composite function block event input.
- Similarly, each event output of a component function block can only be connected to exactly one component function block input event or to one composite function block event output.
- Each composite function block output event must be connected to exactly one output event of a component function block or come directly from a composite function block input event.

Note: Some component function block input and output events may remain unconnected. In this case, algorithms associated solely with unconnected input events will not be executed.

Rules for data connections

The following rules apply to the connections between composite data inputs and outputs and component inputs and outputs.

- Each data input of the composite function block may be (a) connected to one or many data inputs of internal component function blocks, or (b) connected directly through to one or more composite function block outputs or both.
- Each component function block's data input can either be (a) not connected or (b) connected to one other component function block's output or (c) connected to a composite function block's data input. Clearly, the standard does not allow a component function block input to be connected to multiple outputs because the input would then have an indeterminate value.
- Each component function block's data output can be (a) not connected, or (b) connected to one or more component function block data inputs or also (c) connected to one or more composite function block data outputs.
- Each composite function block's data output must be connected to either (a) one component function block data output or (b) 'through routed' from one composite function block data input.

3.5.2 Execution of composite function blocks

Instances of composite function blocks can be created as part of function block networks existing within a resource and may exist at the top level or be used within the definition of other larger composite function blocks. In all situations, the execution of the composite function block is determined by the arrival of events at its event inputs. The standard defines the following simple rules that determine how events are handled:

1. If a composite function block input event is routed through directly to a composite function block event output, an occurrence of an event at the event input will generate an event at the function block's event output.
2. If an input event is connected to an internal component function block, the occurrence of an input event will result in an event arriving at the component function block's input event. The component function block will then be scheduled for execution by the resource's scheduling function.
3. Similarly, if an output event of a component function block is connected to the input event of another component function block, an output event from the first will cause the second function block to be scheduled for execution.
4. If a composite function block's output event is connected to a component function block's output event the composite function block's output event will be generated when the component function block produces an output event.

These simple rules provide an intuitive and logical association between events and function block execution. In essence, events propagate through the network of component function blocks progressing from the input to the output side of the composite function block.

3.5.3 Composite function block example

Consider an example function block with an external interface as shown in Figure 3.12 that has been chosen to demonstrate many of the characteristics of composite function blocks. Note that it has the same appearance as a basic function block. In fact, from an external viewpoint, it is not possible to distinguish between a basic and composite function block just from a function block's external interface.

We will now review the behaviour of this function block and show how it can be modelled as a composite function block. The purpose of this function block is to provide an output value that follows a 'sawtooth' profile as shown in Figure 3.13, i.e.

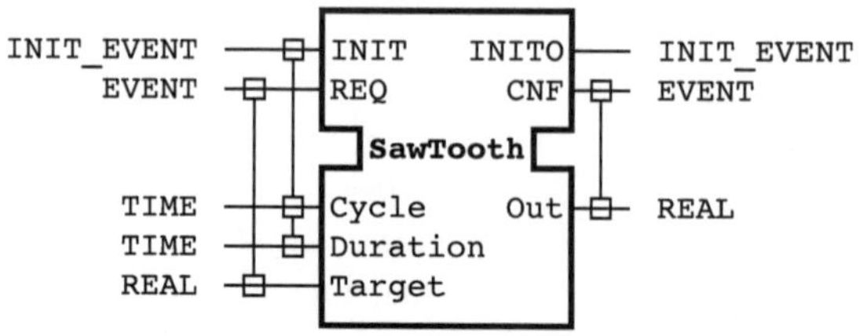

Figure 3.12 Example composite function block – external interface

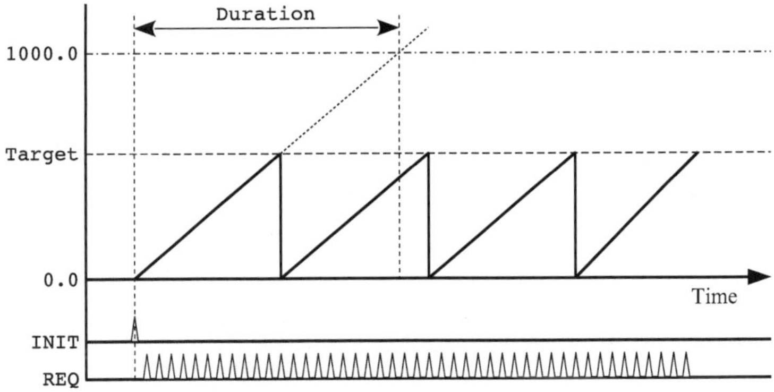

Figure 3.13 Sawtooth profile

that repeatedly ramps the output from 0.0 to a target value and then resets and restarts the ramp again from 0.0.

Figure 3.14 shows the graphical body of a composite function block depicted as a network of function blocks linked together to form a new function block. In this case, an additional component function block of type `Compare` and an Event function block `E_MERGE` have been used to extend the functionality of the `Ramp` function block to form the 'Sawtooth' generator. The instance name of each component function block is shown just above the outline of each function block in the graphical body.

Note: `Ramp1` is an instance of the `Ramp` basic function block as described in the previous section of this chapter.

The function block is initialised using an initialisation event at input event `INIT` – at this point, the input values for inputs `X0`, `X1`, `Cycle`, `Duration` are stored within the `Ramp1` input variables. In this example, `X0` and `X1` are fixed at values `0.0` and `1000.0` by constants defined within the composite function block's body, effectively setting the limits for the `Ramp` output `Out`. The values of `Cycle` and `Duration` come from values provided at the external interface of the function block and define the timing characteristics of the 'sawtooth' waveform.

After initialisation, the function block is ready to receive a regular stream of events at event input `REQ`. Each event at the `REQ` input triggers the `Ramp1` function block to increment its output `Out` towards the upper limit set by `X1`. Each time the `Ramp1` internal algorithm completes its execution, an output event `CNF` is passed to the `Compare1` function block which then checks the value of `Out` against the `Target` value, i.e. it compares inputs `X` and `Y`. After execution, the `Compare1` function block issues event `CNF`, which triggers the `SawTooth`'s `CNF` output event and makes `Ramp1.Out` available at the composite's output `Out`. Eventually, when the `Ramp1.Out` value reaches the `Target` value, the `Compare1` function block detects that input `X` is greater than input `Y` and triggers its output event `E_GT`. This event is fed back to the `Merge1` function block, which produces a new initialisation event for the `Ramp1` function block. The `Ramp1` block is re-initialised and the

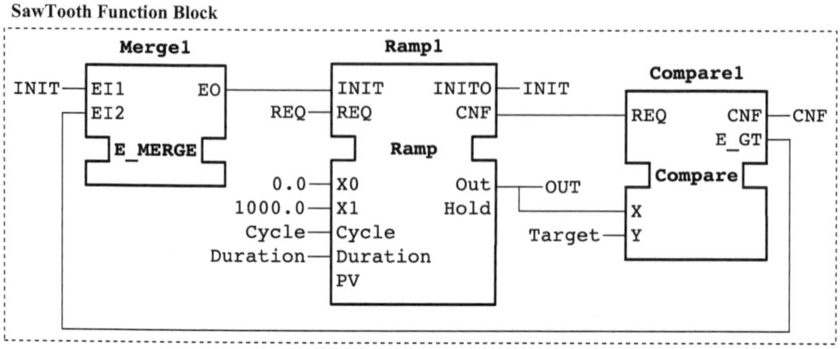

Figure 3.14 Example composite function block – graphical body

output is reset to `0.0`. As long as the function block continues to receive events at its event input `REQ`, it will continue to generate the sawtooth waveform.

The slope of the sawtooth profile can be changed by re-initialising the function block with an initialisation event at event input `INIT` issued with a different `Duration` value. `Merge1` is an instance of one of the standard event function blocks as described in Chapter 6 and provides a means for multiple events to be merged into a new stream of events. In this case, `Merge1` is used to allow the `Ramp1` function block to be initialised either by an external event on event input `INIT` or from an internally generated event coming from the `Compare1` function block.

Textual syntax of composite function block example

So far, we have focused on the graphical representation of a composite function block. As with basic function blocks, the IEC 61499 Textual Syntax can also be used to describe the structure and internal networks that form composite function blocks. For example, the `Sawtooth` function block as depicted in Figures 3.12 and 3.14 can be represented by the following textual description:

```
FUNCTION_BLOCK SawTooth
  (* Event definitions *)
  EVENT_INPUT
    INIT:INIT_EVENT WITH Cycle, Duration;
    REQ WITH Target;
  END_EVENT
  EVENT_OUTPUT
    INITO:INIT_EVENT;
    CNF WITH Out;
  END_EVENT
  (* Variable definitions *)
  VAR_INPUT
    Cycle : TIME;
    Duration : TIME;
```

```
    Target : REAL;
  END_VAR
  VAR_OUTPUT
    Out : REAL;
  END_VAR
  (* Function blocks *)
  FBS
    Ramp1 : Ramp(X0 := 0.0, X1 := 1000.0);
    Compare1 : Compare;
    Merge1 : E_MERGE;
  END_FBS
  (* Event connections *)
  EVENT_CONNECTIONS
    REQ TO Ramp1.REQ;
    INIT TO Merge1.EI1;
    Merge1.EO TO Ramp1.INIT;
    Ramp1.INITO TO INITO;
    Ramp1.CNF TO Compare1.REQ;
    Compare1.CNF TO CNF;
    Compare1.E_GT TO Merge1.EI2;
  END_CONNECTIONS
  (* Data connections *)
  DATA_CONNECTIONS
    Cycle TO Ramp1.Cycle;
    Duration TO Ramp1.Duration;
    Ramp1.Out TO Compare1.X;
    Ramp1.Out TO Out
    Target TO Compare1.Y;
  END_CONNECTIONS
END_FUNCTION_BLOCK;
```

The textual form is rather like a build list for an electronic circuit. It describes all event and data inputs and outputs, the internal function blocks and the event and data connections. Each part of the definition is introduced by a block keyword, e.g. `EVENT_INPUT . . . END_EVENT` defines all input events for the function block, similarly `DATA_CONNECTIONS . . . END_CONNECTIONS` defines all the data connections between the external interface and internal function blocks. In this respect, the definition of the function block network inside the composite function block is similar to the one in applications as shown in the last chapter. The only difference is that within a composite function block external connections can be made to the interface ports provided by the composite function block. An application has no such external interface ports.

The Textual Syntax has been developed to form a generic and portable format that can be used to express the structure of a composite function block. It does not directly express the algorithmic aspects of the function block's behaviour but it does define unambiguously how the function block is constructed, which implicitly defines its behaviour. In fact, software tools can automatically translate between the graphical and textual representations.

3.6 Summary

In this chapter, we have covered most of the important aspects of function block type definition. We have reviewed how type definitions can then be used to create function block instances. In turn, we have seen how function block instances can be used in new type definitions to hierarchically build function blocks of yet higher functionality. Although designs can be created graphically, we have noted that the standard also defines a formal textual syntax that can be used as an alternative representation.

To summarise:

- There are two categories of function blocks: (i) basic and (ii) composite.
- All function block types have an external interface declaration consisting of a static and a dynamic part.
- The static interface defines the event and data inputs and outputs as well as the association of data ports to events using the `WITH` qualifier.
- The dynamic interface description consists of zero or more service sequence diagrams describing the event flow in and out of the function block.
- The interface declaration is all that is required to use a function block in applications or composite function blocks.
- The behaviour of a basic function block is defined in terms of an Execution Control Chart (ECC), one or more algorithms and internal variables.
- Composite function blocks are defined in terms of a network of component function blocks.
- All function block type definitions can either be defined graphically or textually. The graphical and textual representations are interchangeable.

Chapter 4
Structuring applications

In this chapter, we will review how applications can be hierarchically structured into independent sub-systems using subapplications and we will show how the interaction of function blocks can be decoupled by applying the adapter concept.

Specifically we will:

- *review where subapplications can be used and compare their behaviour and properties with those of composite function blocks.*
- *describe the distribution capabilities of subapplications.*
- *consider how the use of subapplications can support either a top-down and bottom-up approach to application design.*
- *decouple application parts and function blocks using common interfaces defined by adapters.*
- *show how adapters can be defined and used in applications and in the different function block types.*

4.1 Application structuring with subapplications

Within an application, further decomposition is possible using **subapplications**. A subapplication has the external characteristics of a function block (i.e. event and data interface), but can contain networks of function blocks that can, themselves, be distributed over other resources. The standard defines a 'fractal' form of application decomposition that allows subapplications to be further decomposed into yet smaller subapplications, if required, as depicted in Figure 4.1. A subapplication is a smaller form of application that can be replicated but otherwise has the same distribution characteristics as an application.

A subapplication can be regarded as a special form of composite function block that is designed to be 'distributed'. That is, it can optionally run on more than one resource. It has a similar structure to a composite function block but some of the rules regarding the use of data and events are relaxed. A subapplication can only be used within the body of other larger subapplications and within applications. However, a subapplication may itself, be defined using composite, basic and service interface function blocks, as well as other subapplications.

The main contrasting feature of a subapplication when compared with a composite function block is that it can optionally be run on multiple resources. Remember that basic and composite function blocks can only run on the same

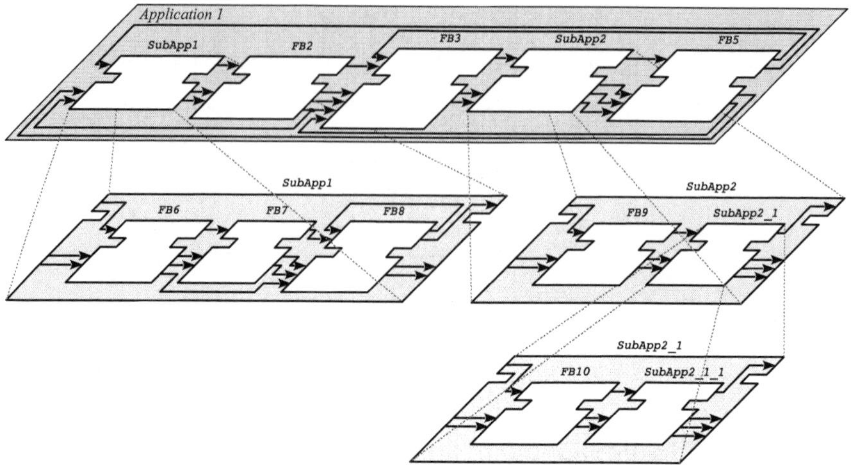

Figure 4.1 Hierarchical application decomposition with subapplications

resource, i.e. it is not possible to split them into parts that can run on different resources. A subapplication, however, may run on one resource or be distributed so that its parts run on different resources; in other words, it is distributable. Extending Table 3.1, Table 4.1 summarises these features and compares them to the other two function block types discussed so far.

Table 4.1 Characteristics of the different function block types and subapplications

Type	Distributable	Definition	Comment
Basic function block	No	States defined using the Execution Control Chart (ECC). Algorithms defined using an appropriate language, e.g. Structured Text, Java.	Basic function blocks define the fundamental function blocks from which large composite function blocks, applications and subapplications can be built.
Composite function block	No	Constructed from a network of function blocks. It is defined by the data and event connections between component function blocks.	A composite function is built from a network of lower level function blocks. These can be either basic or lower level composite or service interface function blocks.
Subapplication	Yes	Constructed from networks of basic, composite and service interface function blocks. A subapplication can in turn contain copies of other subapplications.	A subapplication is intended to provide a reusable part of an application that can be distributed over many resources.

One way of regarding a subapplication is that it represents a reusable section of a function block network. It would typically be used for defining an arrangement of function blocks and connections that can be reused in different network configurations. In many ways a subapplication type definition resembles a software macro in that it allows a particular solution in the form of a function block network to be readily copied.

4.1.1 Rules for subapplication type specification

The rules for subapplication construction are:

1. An instance of a subapplication can only be declared within other subapplication type definitions or within an application.
2. The `WITH` construct is not used in subapplication type definitions because the association between events and data will depend on how the subapplication is distributed between resources.
3. Input and output variables of subapplications are declared in the same way (graphically or textual) as for function block types. However, in contrast to composite function blocks, this does not imply that storage is created for these variables. All inputs and outputs values of subapplications are stored at the interfaces of its component function blocks.
4. Service sequence diagrams can be used for describing the dynamic interface behaviour of subapplications in the same way as for function block type interfaces.
5. The contained function block network consisting of function block and subapplication instances, event connections and data connections is declared in the same way as for composite function blocks.

Figure 4.2 shows these characteristics of subapplications in comparison to composite function blocks.

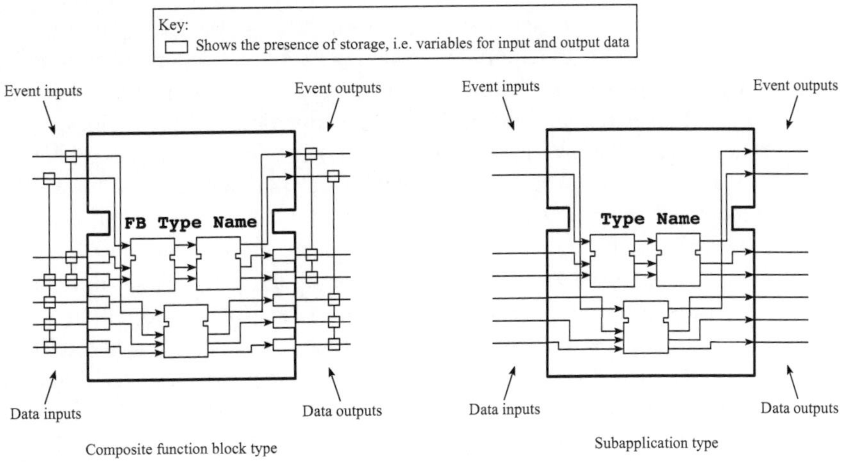

Figure 4.2 Comparison of composite function block with subapplication

4.1.2 Rules for subapplication execution

Instances of subapplications can be declared within applications and also within other subapplications. Nevertheless, an instance of a subapplication is rather different to an instance of a composite function block, in that, it really represents a copy of a set of function blocks and their interconnections. In this regard, a subapplication can be compared to a software macro or a graphical pattern.

The standard defines the following rules for how subapplications execute:

1. Events may be routed directly through the subapplication, so that an event arriving at an event input will immediately trigger an event at the event output.
2. Each event input that is not directly routed through must be connected to an event input of an internal component function block. In this case, when the event arrives at the subapplication event input, the internal component function block will receive an event and will be scheduled for execution.
3. As a component function block executes it will produce one or more output events. These will in turn trigger the execution of other component function blocks to which they are connected within the subapplication.
4. Subapplication event outputs that are not directly routed through from event inputs must be connected to one output event of a component function block. When the associated component function block executes and generates an output event, the event is propagated through to the subapplication's event output.

4.1.3 Subapplication example

For example, consider a subapplication that provides a temperature control loop consisting of an analogue input function block, a PID control function block and an analogue output function block as shown in Figure 4.3. Typically this would be used to control the temperature of a device, for example, a heating vessel or furnace. The primary function of this control loop is to (i) measure the current temperature, (ii) compare its value against the set-point or desired temperature and then (iii) adjust an output value that drives a heating device to correct the temperature. These three functions are mapped onto three component function blocks that make up the subapplication. The `Input1` function block reads the current value of an external sensor. The `PID1` function block provides a PID algorithm to compare the measured (process value) and set-point values and create the output value. The `Output1` function block takes the value created by the PID function block and transfers it to the external actuator.

`Input1` and `Output1` will be constructed as composite function blocks and will each require at least one service interface function block to provide an interface with the underlying controller in order to read input and output values from the controller hardware. Service interface function blocks are a special form of function block that provides various interfaces with the physical device or communications system and are discussed in Chapter 5.

The example subapplication has two event inputs, `INIT` and `Run`. `INIT` is used to initialise the internal component function blocks. The `Input1`

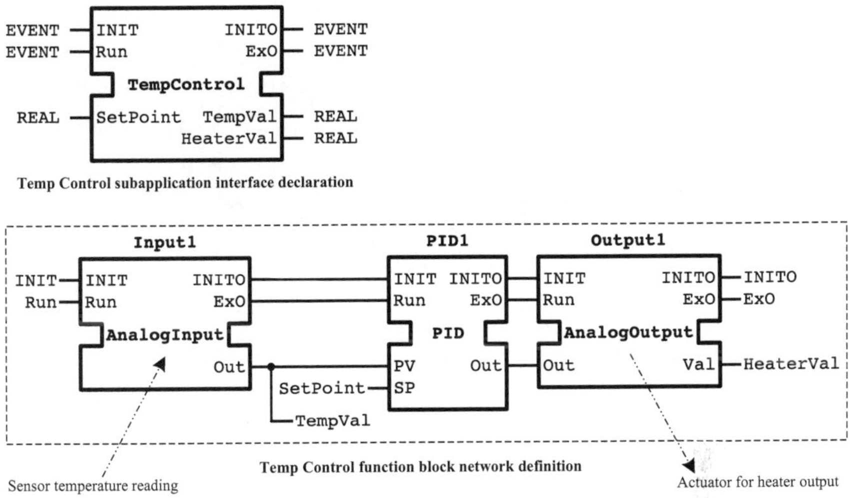

Figure 4.3 Subapplication example: TempControl subapplication definition

and `Output1` function blocks will initialise their respective service interface function blocks. The `INIT` event will also trigger the `PID` function block to read its input variables, i.e. the value of `SetPoint` input is read and stored within the `PID` function block. This defines the desired temperature at which the control loop will stabilise. Generally, a PID algorithm will require a large number of parameters such as timing constants for integral and derivative actions. These are not shown in order to keep the example simple but would be initialised in the same way as the `SetPoint`. The `Run` event propagates through the subapplication and causes the control loop to read the temperature and update its external output according to the value calculated by the PID algorithm. In this example, we can assume that another timing function block within the application provides a regular stream of events, at say, 100 ms intervals at the `Run` event input to ensure that the `TempControl` subapplication is regularly invoked. This will ensure that the temperature is measured at a given scan rate and its value propagated to the `PID1`, which in turn creates a new output value to modulate the heat output.

Figure 4.4 shows how the `TempControl` subapplication could be run within a simple controller to provide closed loop temperature control of a vessel heated using a steam jacket. The input to the `TempControl` subapplication is a temperature value that is read from a sensor such as a thermocouple; the subapplication output drives some form of actuator, such as a steam valve, to modify the heating.

Note: In this example, the `TempControl` subapplication is actually part of the application `SingleVesselControlApplication` that has been assigned to a single IEC 61499 device with a single resource, i.e. all of its function blocks are running within the same resource.

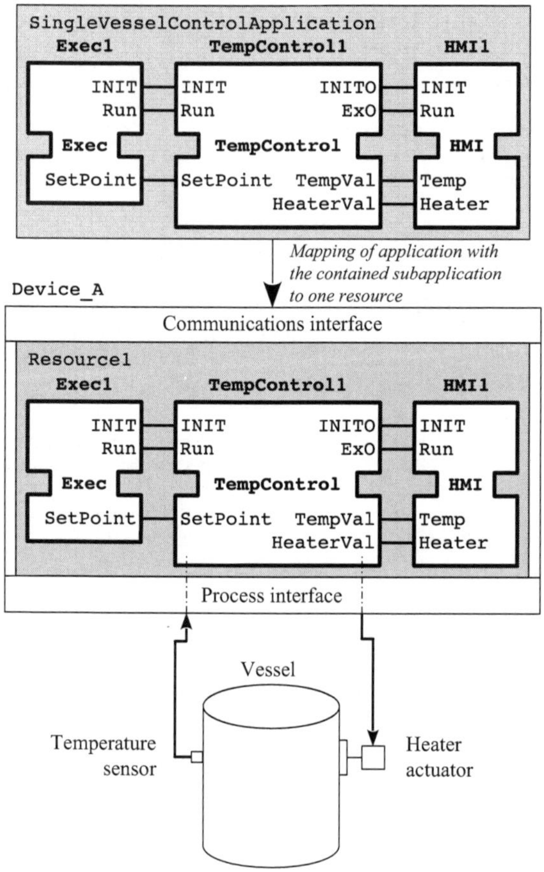

Figure 4.4 Using the TempControl subapplication in an application mapped to a single device

4.1.4 Subapplication distributed example

So far we have considered subapplications that run on a single resource. A significant feature of IEC 61499 is the ability to re-arrange applications and subapplications so that they can provide the same functionality and yet run on different devices and resources in different distribution arrangements. Subapplications such as `TempControl` as depicted in Figure 4.3 can be broken into smaller function block network fragments that run in different resources. In the case of `TempControl` the following arrangements could be considered:

1. All function blocks run on one resource, i.e. a non-distributed configuration as shown in Figure 4.4.
2. Each of the `Input1`, `PID1` and `Output1` function blocks runs on a different resource, i.e. a totally distributed configuration.

3. `Input1` runs on one resource while the function blocks `PID1`, `Output1` run on a second resource – a split resource configuration.
4. The function blocks `Input1`, `PID1` run on one resource while `Output1` runs on a second resource – an alternative split resource configuration.

Figure 4.5 shows one possible physical system configuration that could be considered. It shows two control devices – `Device_A` and `Device_B` – linked by some form of communications system, e.g. a network using Fieldbus or Ethernet. The analogue input is connected to `Device_A` while the actuator is now driven from the other `Device_B`. The subapplication is now distributed over two controllers but the internal functionally is still determined by the function blocks defined in the original `TempControl` subapplication type definition as given in Figure 4.3.

Utilising this system configuration, Figure 4.6 shows the fourth distribution arrangement where the function blocks `Input1` and `PID1` for the analogue input and PID algorithm run on one resource `Resource1` in `Device_A` and the analogue output `Output1` runs on a second resource `Resource1` in `Device_B`. In practical terms, this arrangement means that the subapplication `TempControl` can be run in two separate resources located in different devices.

It is envisaged that a system designer will be free to select the appropriate distribution arrangement for a subapplication according to the system design requirements. The distribution arrangement selected by a designer will depend on many factors including: (i) whether a resource is capable of supporting particular types of function blocks, (ii) the loading and performance of a particular resource and (iii) the latency and reliability of communications services between resources.

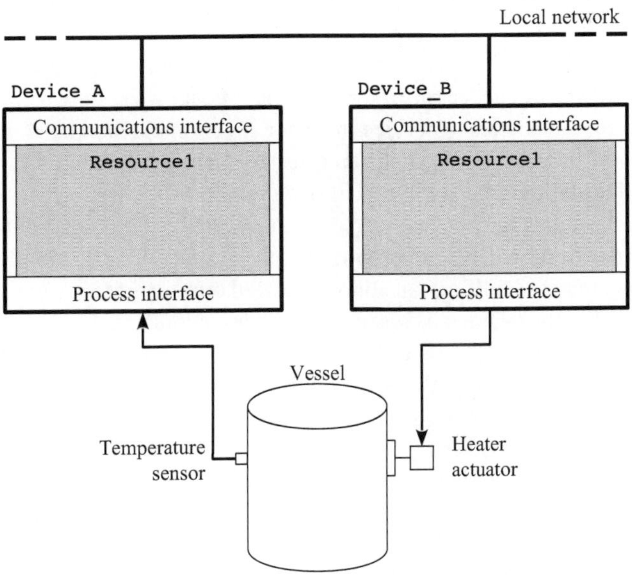

Figure 4.5 System configuration for distributed temperature controller

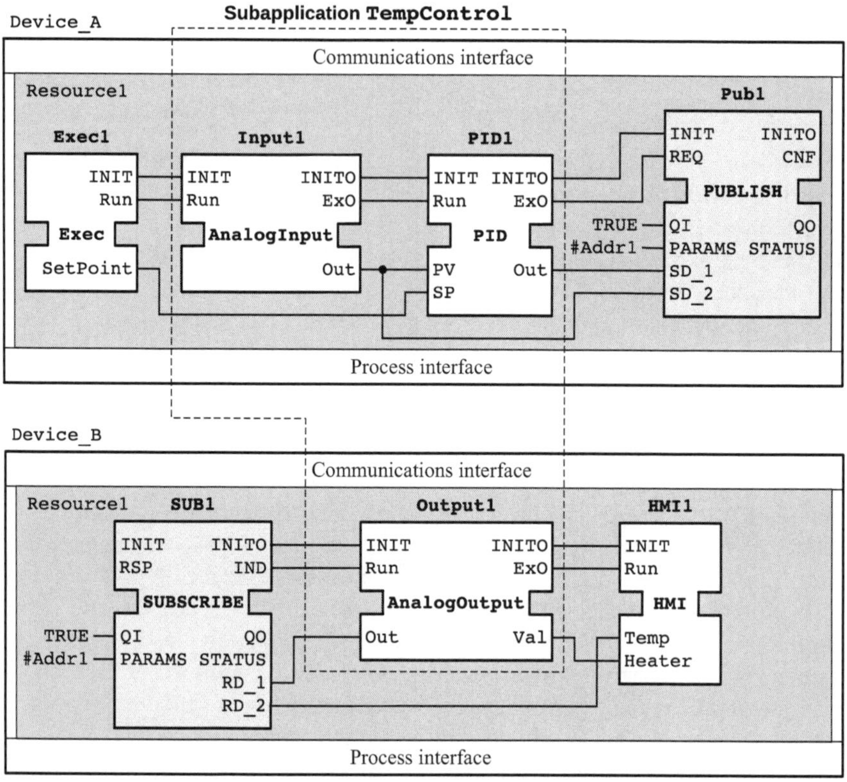

Figure 4.6 TempControl subapplication distributed example

It will be noted that two extra function blocks `Pub1` and `Sub1` have been added to the distributed `TempControl` subapplication as shown in Figure 4.6. These are further examples of service interface function blocks that are discussed in more detail in Chapter 5 and provide an interface between function blocks within a resource and the resource's communications system. `PUB1` is a `PUBLISH` function block that is used to transmit one or more data values over a communications link to one or more external resources within other controllers. It simply sends out values using a given identification address (e.g. in our example `#Addr1`). `SUB1`, a `SUBSCRIBE` function block is used within the second resource to receive the values sent out by `PUB1`. In this case, there is only one subscriber, but it is possible to have configurations where there are multiple subscribers.

The example has used the `PUBLISH/SUBSCRIBE` service interface function blocks but other blocks offering different types of communications models, such as Client/Server, are also discussed in the standard (see Section 5.4). In this particular example, the Publisher/Subscriber model has been used because there is no requirement in this simple design for any feedback to the `PID1` function block as to whether the transmission of the value to the analogue output has been successful. The `PID1` function block will continue to output new values to its output regardless of downstream communications problems. Clearly, in a full design there would be a requirement to monitor the communications and check whether the complete control loop is working correctly. This could be achieved by using, say, a Client/Server communication model or by feeding back a signal from the `Output1` function block.

Note: When a subapplication is re-arranged to run on different resources, additional parameters will generally be needed for the communications system. This would typically include details such as resource addresses and network routing parameters (as discussed in Section 2.4). Such information will generally be passed as parameters to service interface function blocks that are handling the inter-resource communications. It should be stressed that IEC 61499 does not define communications protocols or standardise addressing or parameters related to communications. But it does provide a framework and architecture for defining services such as communications using service interface function blocks, which will be described in the next chapter in detail.

4.1.5 Guidelines for using subapplications

As subapplications are very similar to composite function blocks it is not always clear when to use them. Composite function blocks encapsulate and hide their implementation in an atomic way (i.e. the internal design is not visible to and distributable by its user). Therefore composite function blocks are an appropriate means when functionality needs to be encapsulated in a reusable way and the user of the functionality does not need any information on the internals. In contrast subapplications provide a function block network in a reusable but open way. This supports instance specific adjustments (i.e. parameters) according to the needs of the application or the system configuration.

The following rules of thumb summarise when to use a subapplication instead of a composite function block:

- Other subapplications are to be used.
- Application or system configuration specific parameters and adjustments are needed but it is not appropriate or efficient to provide them at the outer most interface of the container. For example, process parameters of a sub-system which are provided by the specific process engineer in charge of this sub-system. In other words, the application specific parameters are embedded within the subapplication.
- The function block network has the potential or requirement to be distributed to different resources during the mapping stage. An indicator that this is the case is that two or more service interface function blocks are used, as the services encapsulated by them could be provided by different devices.

During the application development process subapplications can be defined and used in a top-down, bottom-up or mixed approach. Pursuing a top-down approach implies starting with an empty application and then drafting subapplications and their interfaces forming a hierarchy of application parts according to the requirements of the application. Identified subapplications can be developed subsequently and, if required, by different independent development teams. In a bottom-up approach existing subapplications or newly defined subapplications together with function blocks would be used to form new applications and sub-applications. Considering existing subapplications in the decomposition phase of the top-down approach results in a mixed design approach.

4.2 Structured interfaces using adapters

One of the biggest points of criticism in IEC 61499 has been that the event and data interface is separated. Although this separation is advantageous in use cases where event and data has different origins it leads in general to cluttered design spaces. Especially when modelling the interaction of two independent application parts as shown for example in Figure 4.7. This shows two function blocks that perform some handshake interactions where `FB1` is sending one event and three data points to `FB2`, which sends two events and one data point back to `FB1`.

Apart from the event and data connections cluttering the design space this example shows additional problems of modelling the interaction of these two function blocks in this way:

- The interaction of the two function blocks is not clearly visible. Furthermore there is no element which describes the interaction semantics (i.e. in which order events have to be sent between both parties).
- Misusing the interface is very easy. The user may omit to connect all the required interface elements or wrongly connect interface elements (e.g. interchange two connections).

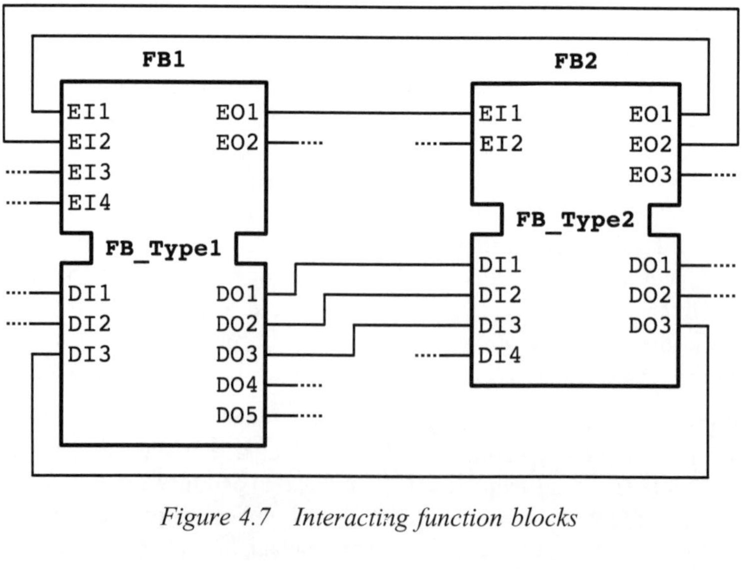

Figure 4.7 Interacting function blocks

4.2.1 Adapter concept

In order to solve these issues IEC 61499 provides the adapter concept which supports the capture and combination of such interactions into one clearly defined interface. Applying the adapter concept reduces the interaction to a single interface element on both sides with a single connection between them. An adapter can be used to connect data and events, i.e. any number of designated inputs, outputs and associated input and output events that need to be passed between two function blocks. The names of the inputs and outputs (for data and events) are specified in the adapter type specification. It should be noted that an adapter type has no algorithms, state or constants. It is just an extension of the function block's interface. However data and events that are provided via an adapter are not exposed on the function block interface. They are only usable within the function block and cannot be used in graphical connections to other function blocks. The adapter concept can therefore be used to improve 'information hiding' and help to simplify the top-level design.

An adapter can be regarded as an analogy of an electrical cable which connects function blocks via a 'plug' and a 'socket'. A plug is used in the function block that is regarded as 'providing' the data and events; the socket is used in the function that is regarded as 'requesting' the data and events. The transformation from the event and data interface, as applied in Figure 4.7, into an interface utilising the adapter concept is depicted in Figure 4.8.

Function blocks may have more than one adapter interface to provide dedicated interfaces to a number of different function blocks. For example, we could envisage a function block handling an analogue input having separate adapter interfaces to function blocks for alarm management and calibration.

Note: It is not possible for adapter interfaces to be unconnected. Any function block that has been specified with either an adapter plug or socket must have an adapter connection to an appropriate function block providing the opposite adapter interface.

Summarising the core concepts of adapters are:

- An adapter connection combines the bidirectional interaction of two function blocks encapsulating several events and data connections in both directions.
- Adapter ports are marked with '>>' in the function block interface.
- Function blocks can define adapter ports in the form of a requesting interface. These are called Sockets in IEC 61499 and are depicted at the bottom left (input side) of the function block. Requesting interface means that the function block requires an implementation for this interface.
- Function blocks can define adapter ports in the form of a providing interface. These are called Plugs in IEC 61499 and are depicted at the bottom right (output side) of the function block. Providing interface means that the function block provides an implementation for this interface.
- An adapter connection always connects one plug with one socket of the same adapter type.

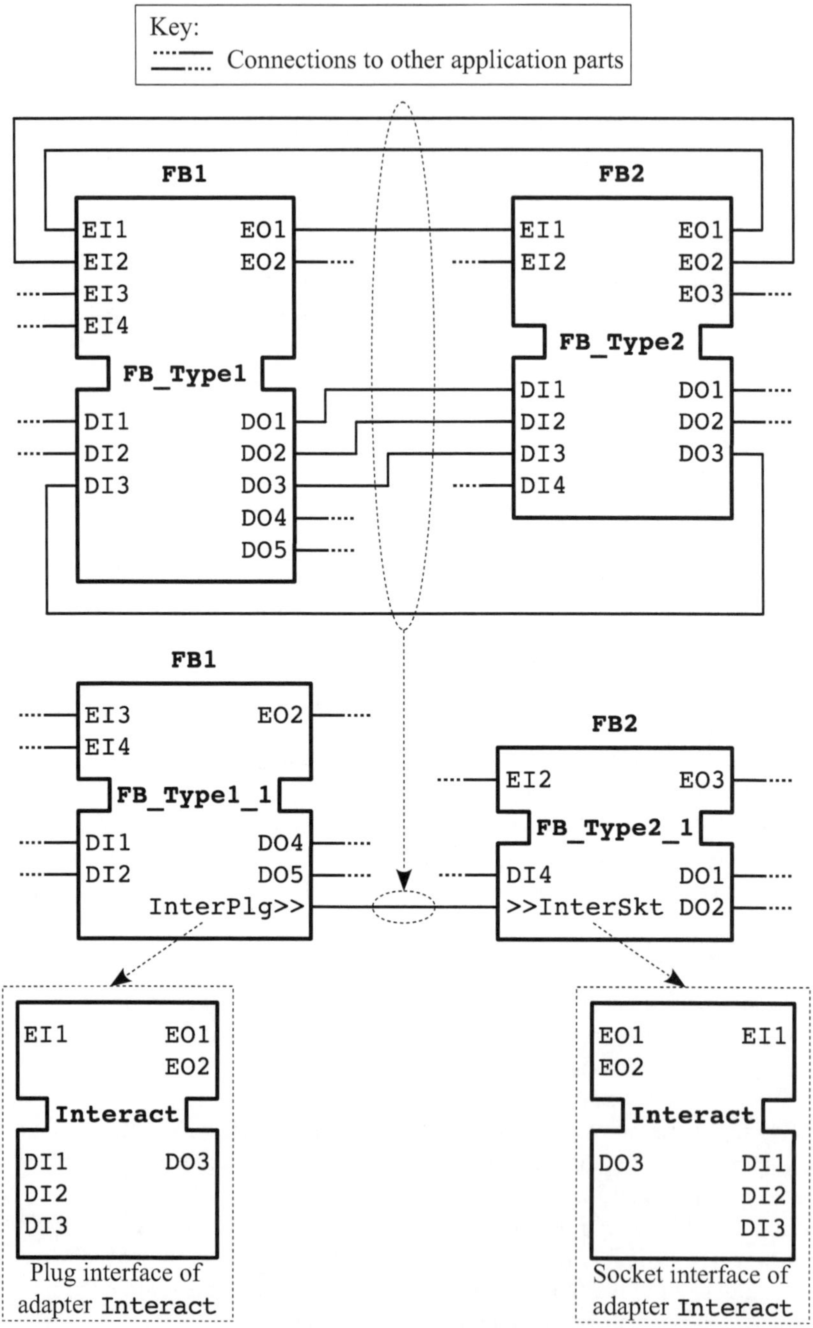

Figure 4.8 Representing structured interfaces with adapters

- The interface of adapter types is defined in the form of a function block interface.
- The interface of plugs and sockets mirror each other. That means events and data entering on either plug or socket leave on the corresponding other end of the connection. For example triggering the event `EI1` of `FB1`'s plug `InterPlg` will result in the reception of the `EI1` event at `FB2`'s socket `InterSkt`.

Note: From the software modelling point of view the adapter concept corresponds closely to the port concept as it is applied in ROOM [23] or UML 2.0 [22].

4.2.2 Defining adapter types

The interface of adapter types is defined in the same way as for a function block. It consists of the type name, the event and data interfaces and the assignment of input and output variables to input and output events respectively using the `WITH` construct. The interface declaration can be performed either graphically or textually (i.e. using either textual syntax or XML format). The interface is declared and stored in the exchange format from the point of view of the requesting side of the adapter interface (i.e. the socket). However tools may provide means which allow the declaration of adapter interface types either from the plug or socket point of view. Figure 4.9 depicts such an adapter interface declaration for the example adapter type `Interact` from Figure 4.8.

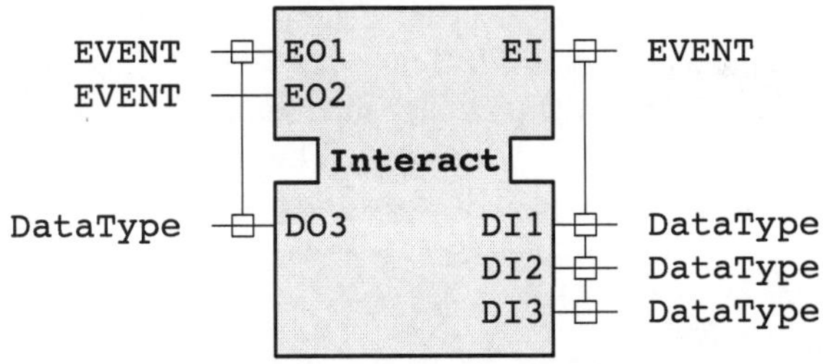

Figure 4.9 Graphical adapter type interface specification

Note: In IEC 61499 there is no way to distinguish between function blocks and adapters when only viewing the interface. In figures in this book, to clarify where adapters are being used, they are shown with grey shading although this is not a requirement of IEC 61499.

The textual syntax to describe an adapter interface is as follows:

```
ADAPTER Interact
  (* Event definitions *)
  EVENT_INPUT
    EO1 WITH DO3;
    EO2;
  END_EVENT
```

```
  EVENT_OUTPUT
    EI WITH DI1, DI2, DI3;
  END_EVENT
  (* Variable definitions *)
  VAR_INPUT
    DO3 : DataType;
  END_VAR
  VAR_OUTPUT
    DI1 : DataType;
    DI2 : DataType;
    DI3 : DataType;
  END_VAR
END_ADAPTER
```

The main difference from a declaration of a function block is that for adapters, the adapter declaration is enclosed with the keyword ADAPTER … END_ADAPTER.

4.2.3 Modelling the behaviour of adapter types

In order to describe the dynamic behaviour of adapter types IEC 61499 allows the definition of service sequence diagrams for adapter types as for normal function blocks. In contrast to service sequence diagrams for function blocks, the service sequence diagrams define the interaction of the plug and the socket. Therefore, the interface elements in the service sequence diagram are depicted with the plug on the left and the socket on the right side (Figure 4.10). As events entering either plug or socket are directly forwarded to the other end of the adapter connection, the service sequences are always symmetric. The main task of service sequence diagrams for adapter interfaces is to describe the event order and which element – plug or socket – is initiating a transaction.

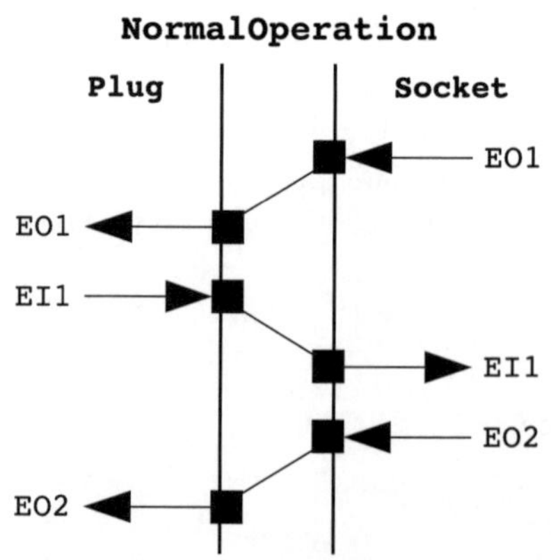

Figure 4.10 Describing the behaviour of adapter interfaces using service sequence diagrams

4.2.4 Using adapter types in function blocks

As stated above, adapter type instances can be used in function blocks either as plugs or as sockets. In both cases the instance is identified with an instance name. In the graphical representation sockets are declared and shown at the bottom left – the input side – of the function block interface. In the textual syntax sockets are declared within a `SOCKETS ... END_SOCKETS` block. Plugs are declared and shown in the graphical representation at the bottom right – the output side – of the function block interface. In the textual syntax plugs are declared within a `PLUGS ... END_PLUGS` block. In order to differentiate plugs and sockets from data inputs and outputs they are denoted in the graphical representation with a '≫' symbol.

Within a function block type declaration, the inputs and outputs of plugs and sockets shall be used in the same manner as the inputs and outputs of the function block. However it should be noted that semantically the outputs of a plug or socket act as input to the function block, as these are the events and data received from the other end of the adapter connection. The inputs of a plug or socket act like outputs of a function block as these are the events and data to be sent to the other end of the adapter connection.

Adapters may be used in any function block type (i.e. basic, composite, or service interface). The specific usage of the inputs and outputs of plugs and sockets depends on the function block type in which it is used and is described as follows – along with examples of the associated textual syntax.

Adapters in composite function blocks

The most intuitive way of using adapters is within composite function blocks. The reason for this is that the adapter interface of plugs and sockets declared in composite function blocks is shown in the internal function block network of the composite function block. Figure 4.11 shows this for the example function block `FB_Type2_1` used in Figure 4.8. Inputs and outputs of plugs or sockets can be easily used in composite function blocks by using them in connections between internal function blocks. In the textual syntax the adapter instance name of the plug or socket together with the interface element name is used for identifying the input or output (e.g. `InterSkt.EI1`). This is similar to using a component function block.

Adapters in basic function blocks

Within basic function blocks the inputs and outputs of plugs and sockets can be used in the Execution Control Chart (ECC) and in algorithms. In the ECC output events (i.e. the triggers from the other end of the adapter connection) of plugs and sockets can be used for the transition conditions, while input events (i.e. notifying the other end of the adapter connection) of plugs and sockets can be used in the action statement of a state. Data inputs and outputs of plugs and sockets can be used in the transition conditions of the ECC and within algorithms. In order to distinguish the inputs and outputs of plugs and sockets from the inputs and outputs of the function block, the adapter instance name is again used as the identifier. Figure 4.12 shows a part of an ECC and an algorithm utilising inputs and outputs of the plug with the instance name `InterPlg`.

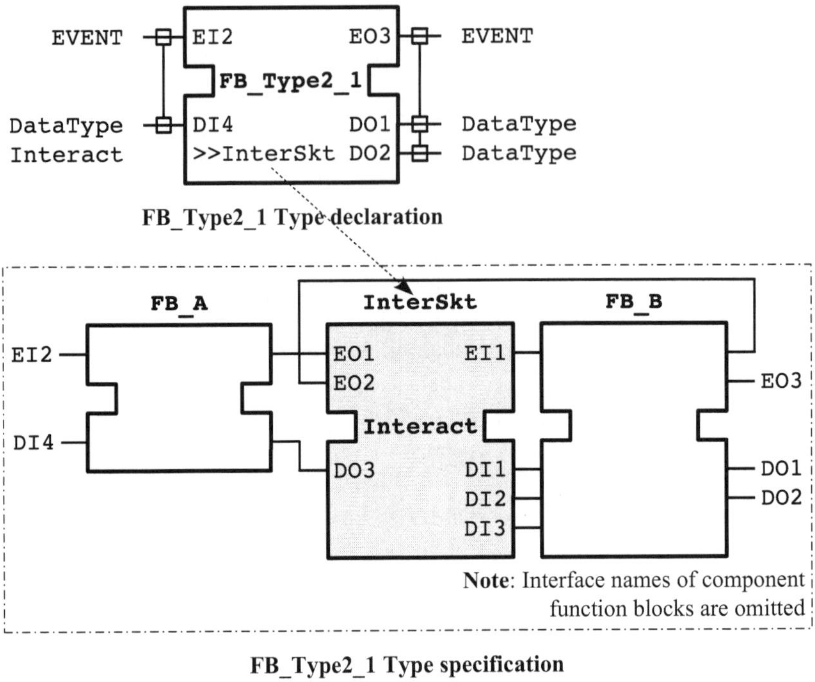

Figure 4.11 Using adapters in a composite function block type

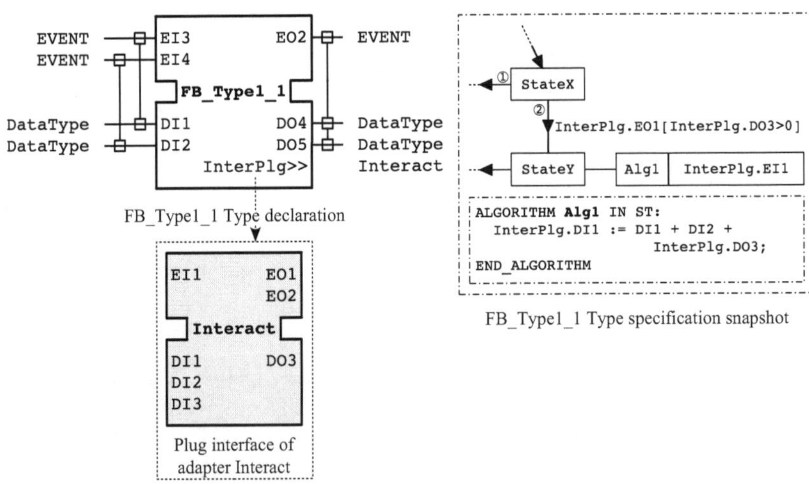

Figure 4.12 Using adapters in a basic function block type

Adapters in service interface function blocks

The usage of adapters within service interface function block is not defined within the standard. However inputs and outputs of plugs and sockets can be used in service sequence diagrams in a similar way to the use of inputs and outputs of

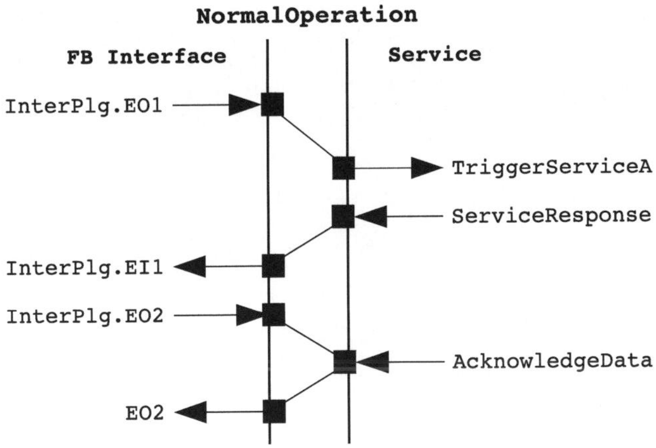

Figure 4.13 Describing adapter service interface function block interaction with service sequence diagrams

function blocks (Figure 4.13). Again the adapter instance name is used as identifier for the inputs and outputs of plugs and sockets. Furthermore this shows how service primitives using adapter events and normal function block interface events can both be used in a service sequence diagram.

For implementing service interface function blocks, the implementation platform specific means for handling events and data of adapters should be made available, this is similar to the way used for handling events and data of the service interface function block itself.

4.2.5 Adapter interface usage example

Consider for example a sub-system in a machine where parts are delivered on a conveyor and taken by a pick and place unit from the conveyor in order to be processed, as depicted in Figure 4.14. Furthermore the conveyor is able to take back parts provided by the pick and place unit and transport it back; that is, parts removed can be replaced on the conveyor. The pick and place unit considered in this example consists of two linear axes denoted in Figure 4.14 with X and Y and a gripper. For this example we will not consider the control of the conveyor or the pick and place unit but focus on how to model the interaction of both parts' control application using an adapter interface.

For modelling the adapter interface of the interaction of the conveyor with the pick and place unit, we will consider that the conveyor provides the interface (i.e. the plug) and the pick and place unit requires the interface (i.e. the socket). The interaction will take place in such a way that the conveyor will inform the pick and place unit if there is a part ready to be taken or that the conveyor is free to receive parts. When a part is ready to be taken an identification number is provided by the conveyor. The pick and place unit can report to the conveyor that the part has been picked from or that a part has been placed on the conveyor. In the latter case, the pick and place unit also provides an identification number

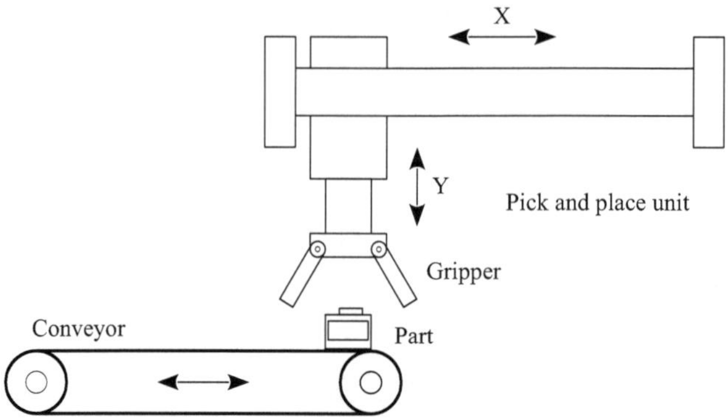

Figure 4.14 Pick and place unit handling parts on a conveyor

of the part that has been placed on the conveyor. Figure 4.15 shows the type definition of the `ConveyorInterface` adapter interface defining this interaction behaviour.

Note: This interaction protocol is rather simple. For a real-world use case the error interaction (e.g. part could not be picked) would also be required.

Using the `ConveyorInterface` adapter is pretty straight forward. Consider two composite function blocks or subapplications which are implementing the control of the conveyor and the pick and place unit. Both would internally use service interface function blocks for interacting with the sensors and actuators and basic function blocks for providing the control functionality (e.g. the movement sequence of the pick and place unit). These component function blocks would interact with the socket `Conveyor` of the `PickAndPlace` function block on the

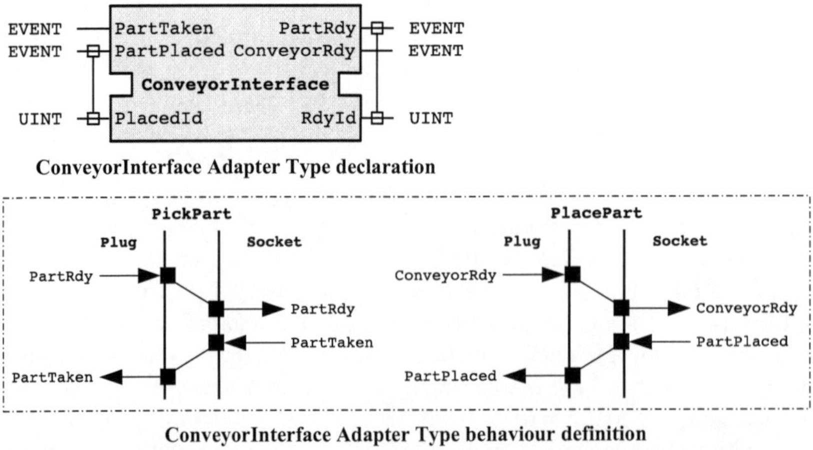

Figure 4.15 ConveyorInterface adapter type definition

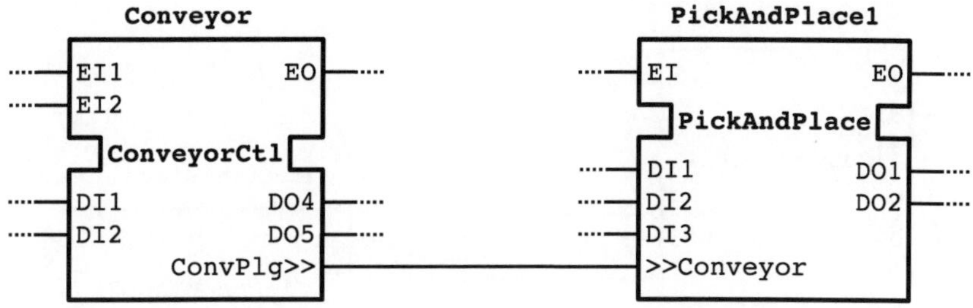

Figure 4.16 Using the ConveyorInterface adapter

one side and with the plug `ConvPlg` of the `ConveyorCtl` function block on the other side, as shown in Figure 4.16.

So far this example mainly shows the advantage of the reduced wiring effort (i.e. one connection versus six in the non-adapter case). However the main advantages of adapters can be gained when applications have to be changed or application parts have to be reused in a different scenario. Consider for example the extension of our pick and place scenario with a second conveyor, as depicted in Figure 4.17. Now the pick and place unit is in charge of transferring parts between the two conveyors.

For modelling the control application for this new system we can directly reuse the `ConveyorCtl` function block developed for the first scenario without any changes. The only new development that is required is a new function block for coordinating the pick and place unit. This function block will now get two sockets one for each conveyor. This provides an easy way to model the additional interactions between the two conveyors. The resulting control application for the second scenario with the two conveyors is depicted in Figure 4.18.

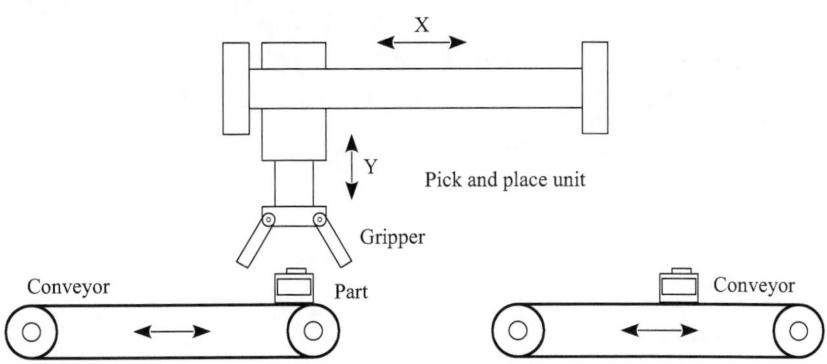

Figure 4.17 Using a pick and place unit for transferring parts between two conveyors

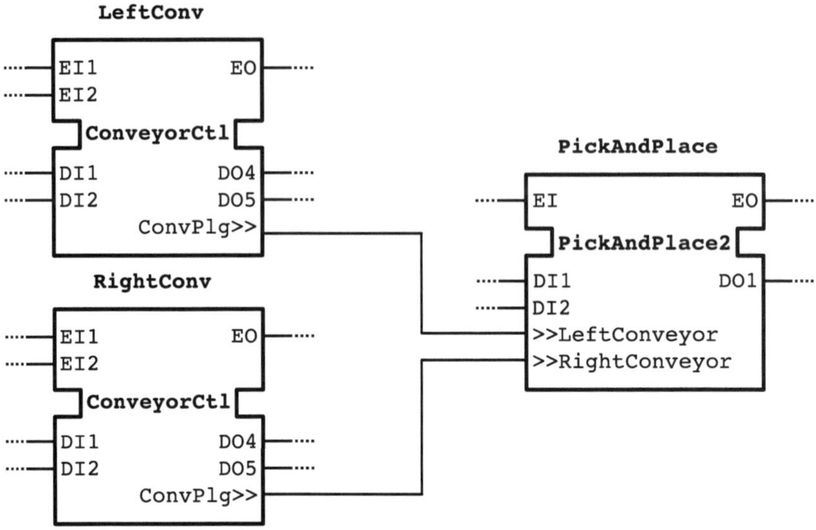

Figure 4.18 Using the ConveyorInterface adapter for interacting with two conveyors

4.2.6 Guidelines for using adapters

Adapter interfaces can greatly improve application design and maintainability. However the following points should be considered when using and developing adapter interfaces:

- An adapter interface describes the interaction protocol of two entities for a certain use-case. Therefore, the adapter interface definition should only contain the elements required for this use-case. Putting implementation specific elements into the interface makes it harder to replace the plug or socket provided by the function block.
- Adapter interfaces describe a bidirectional set of events and data which belong together. When developing adapter interfaces based on existing application designs not all events and data of interacting application parts should be put into one adapter. Better group events and data according to the interaction principle. This results in more general reusable adapter interfaces.
- Using an adapter interface declares a decoupling and design variability point. That means that the function block using the adapter interface is not interested in the specific implementation and can be flexibly used with any function block that provides the other end of the adapter interface.

4.3 Summary

In this chapter we have covered how to improve application designs with the means for structuring and decoupling defined in IEC 61499. The main element for structuring applications is the subapplication. In turn, we have seen that subapplications provide a means of encapsulation that resembles composite function blocks while at the same time allowing application specific configuration and distribution to several devices. We have reviewed how adapter interfaces can provide structured interfaces for function blocks and how they can reduce the number of connections. Furthermore, adapter interfaces support the decoupling of independent interacting application parts into independently maintainable elements.

To summarise:

- A subapplication can be designed so that it can be distributed to run on one or more devices.
- Subapplications allow the decomposition of applications in a hierarchical manner.
- Subapplications are similar to composite function blocks but also have the flexible characteristic that they can be distributed to run on more than one resource.
- Use of subapplications supports either a top-down or bottom-up approach to application design.
- Adapter interfaces allow the composition of bidirectional interaction of two function blocks consisting of several event and data points.
- Function blocks can utilise an adapter interface either as the providing interface – designated as a 'plug' – or as the requesting interface – designated as a 'sockets'.
- Adapter interfaces can be used in any function block type:
    - In basic function blocks adapter inputs and outputs can be used in the transition conditions and state action statements of the execution control chart and in algorithms using the adapter instance name as identifier.
    - In composite function blocks the inputs and outputs of adapters can be used as connection endpoints in the contained function block network.
- Adapter interfaces decouple application parts so that it is possible to change and maintain both sides independently.

Chapter 5
Service interface function blocks

This chapter reviews a special form of function block that provides interfaces into the underlying resource and communications systems.

Specifically we will:

- *discuss why service interface function blocks are required and show where they can be used,*
- *review standard input and output data and event points required in service interface function block type definitions,*
- *review how service sequence diagrams are used to describe the behaviour of service interface function blocks,*
- *consider some examples of service interface function blocks and look at where they could be applied,*
- *describe the two communication models defined in the standard and review the corresponding communication service interface function blocks,*
- *finally we will review Management function blocks, which are a further specialised form of service interface function block for controlling the creation and management of function blocks within resources and devices.*

5.1 Overview

So far we have focussed on function blocks used to model the system independent behaviour of an application and how to model the behaviour of control functionality with basic and composite function blocks. In Chapter 2, we reviewed the creation of a simple distributed application and showed that this was only possible if additional **Service Interface (SI) function blocks** were provided to communicate data values and events between resources. In fact, wherever there is any form of interaction required between function blocks within a resource and the external world, there is a requirement for an SI function block. IEC 61499 does not standardise on particular types of SI function blocks but does stipulate that these forms of function block should be defined using a standard set of input and output variables and input and output events. There is also a special notation used to describe the sequence of interactions, such as, sending a request and waiting for a response that occur externally during the execution of an SI function block.

We will now consider where we might need SI function blocks. In an industrial controller there is clearly a requirement to read the values of physical inputs such as those from pressure and temperature sensors and also to write output values to actuators for example, for driving devices such as valves, pumps, motors and so on. There are also requirements to transmit values over serial communications links, to send out copies of data to external controllers, drive displays and read inputs from display panels and other HMI devices. In fact, these are all examples of external interactions with a resource that can be modelled using different types of SI function blocks.

Some examples of SI function blocks that might be used to model an industrial control system are set out in Table 5.1. These are just a few typical examples of the kinds of SI function blocks that might be required. It should be noted that the standard does not attempt to define specific SI function blocks as is it clear that every system will have its own particular special requirements. However, we will see that IEC 61499 does standardise some important aspects of the SI function block interface including the main input and output parameters as well as the generic behaviour of certain kinds of SI function blocks.

Table 5.1 Service interface function block examples

Type name	Service provided	Application example
IO_Writer	Allows a resource to transmit one or more values to locally connected I/O device(s). **Note:** IEC 61499 assumes that the controller I/O sub-system lies outside the resource but can be accessed using SI function blocks (e.g. I/O ports or I/O modules connected via a backplane bus).	Used to update the value of actuators driving physical devices, such as, valves, heaters or pumps.
IO_Reader	Allows a resource to receive the latest value(s) read in from one or more devices.	Read the latest values of positions of input devices, e.g. micro-switches on a robotic placement mechanism.
Publisher	Transmits the value(s) of one or more variables to one or more external resources that support a 'subscriber' service.	To continuously transmit output values to a number of remote resources, e.g. to update values for controllers that drive actuators.
Subscriber	Receive one or more values from an external resource that offers the publisher service. **Note:** The publisher and subscriber will identify a particular set of values by a network specific unique address or service identifier. There can be many subscribers receiving data from a single publisher.	To regularly receive a stream of values from a given external resource, e.g. to update a set of HMI displays.

5.2 Interface specifications

SI function blocks provide external services to applications. Unlike the other function block types, they cannot be fully modelled using IEC 61499. Only the interface of the function block providing the service can be modelled. Therefore the interface specification constructs for SI function blocks have been further refined for simplifying this task.

5.2.1 Standard inputs and outputs for service interface function blocks

Every SI function block should use the following standard inputs and outputs, although not all of these will be used necessarily in every SI type definition:

Event inputs

`INIT` This input event is used to initialise a particular service that has been provided by the function block. For example, it could start a service to provide data transmission over a serial link. This event is typically sent with a number of input parameters to characterise the type of service such as, e.g., network address and Baud rate.

`REQ` This event initiates a request to obtain data from an external agent. For example, this event could be used to initiate a transmission of a request to get data from an external device.

`RSP` This event initiates the transmission of a response to an external agent. For example, it could send data to a remote HMI device in response to a request for data.

Event outputs

`INITO` This output event signals that the SI function block has completed its initialisation, i.e. as a result of receiving an `INIT` event. It does not necessarily mean that the service has been initialised successfully; a `STATUS` output is provided for this purpose.

`CNF` The 'confirmation' event is issued when the function block has completed the transmission of a request to an external agent. For example, it should be used to indicate that a request to read a particular physical I/O point has been processed by the controller's I/O sub-system.

`IND` The 'indication' event is output when the SI function block has received a response from an external agent. For example, `IND` event would be produced when an I/O read from the controller's I/O sub-system has obtained the value from the selected sensor.

Data inputs

`QI : BOOL` The `QI` data input is used with input events as a simple qualifier input (i.e. enabling or disabling a service execution).

For example, when used with an initialisation `INIT` event, `TRUE` indicates that the service provided by the function block should be initiated, and the value `FALSE` indicates that the service should be terminated.

`PARAMS : ANY` This input represents a data structure that holds a set of values that are associated with the SI's service and define its particular characteristics. The data types and number of values within the structure will be specific to the type of service provided by the function block. The SI function block type definition will define the `PARAMS` structure and default values of parameters within it. This input is only used with the `INIT` event and is used for the initialisation of the service.

For example, a `PARAMS` input for a communication SI function block would contain network addressing information and other communications characteristics, or when writing values to output devices, these parameter values will contain the hardware addresses (such as rack, module, channel).

`SD_1, ... SD_m : ANY` These data inputs are used to send data with requests and with responses. The number of inputs and their data types will be specific to the type of service provided by the function block; this is shown by the notation `SD_1, ... SD_m`.

For example, when writing values to output devices, these variables will contain the output values to be written.

Data outputs

`QO : BOOL` This qualifier output is used to indicate whether the service has been successfully completed for any input event. For example, following an initialisation `INIT` event, a `TRUE` value would indicate a successful start-up; `FALSE` would indicate that the service has failed to be initialised.

`STATUS : ANY` This output can be set with any of the input events and is used to provide the status from processing the last input event. For example, status will be set when a service is initialised using the `INIT` event and has been unsuccessful. Similarly when an `REQ` event used to transmit values to a remote device is processed and subsequently fails, `STATUS` is used to hold the reason for the failure.

`RD_1, ... RD_n : ANY` These outputs are used to convey data received from confirmations and indications. As with the `SD_1, ... SD_m` inputs, the number of these outputs and their data types will be specific to the type of service being provided.

For example, consider when a request has been made to read, say, the values from a set of 10 input points – an `CNF` event will occur when the service has received the values from the I/O sub-system; their values will be available on a set of `RD_1, ... RD_10` outputs.

5.2.2 Service sequence diagrams for service interface function blocks

Apart from the static interface, the service sequence diagrams are the only means for describing the dynamic behaviour for SI function blocks (i.e. how it will behave

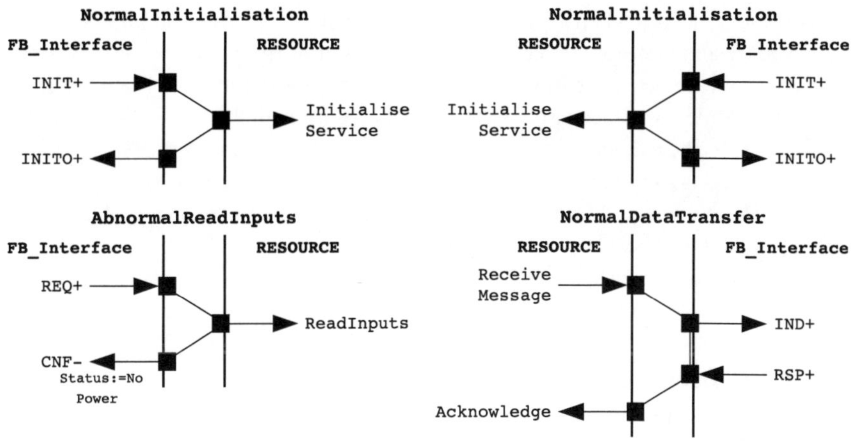

Figure 5.1 Describing service interface function block behaviour with service sequence diagrams

as well as how it should be used). For SI function blocks, the two sides of the interface used in a service sequence diagram (see Section 3.3) are the function block interface – denoted by the function block type name – and the resource interface. For SI function blocks which are mainly triggered by the application (e.g. a function block reading inputs) it is the convention to put the function block interface on the left and the resource on the right side (left column in Figure 5.1). For SI function blocks where the resource is the main trigger (e.g. receiving messages from a network) it is the convention to put the resource on the left and the interface on the right side (right column in Figure 5.1).

The event name suffix '+' indicates whether the event is associated with a successful (or normal) transaction while the suffix '–' is associated with an unsuccessful (or abnormal) transaction. With input events, the suffix '+' indicates that the value of the input `QI` that is set with the event will be `TRUE`. Conversely, the suffix '–' implies that the input `QI` will be set to `FALSE`.

For example, the event `INIT+` indicates that it will be received with `QI` set to `TRUE` and implies that the service should be initialised. In contrast, `INIT–` indicates that the event will be received with `QI` set to `FALSE` and signals that the service should be terminated.

Similarly with output events, the suffix '+' indicates a successful or positive event while the suffix '–' indicates that the event is associated with an unsuccessful or negative transaction. With output events, a '+' suffix indicates that the value of output `QO` will be set to `TRUE`, while a '–' suffix indicates that `QO` will be set to `FALSE`.

For example, `CNF–` would be a negative confirmation of a service request and would imply that transaction has in some way been unsuccessful. If this occurs in response to a request it could either mean that the data provided is not valid or that the underlying service has an issue. Typically, the `STATUS` output gives further information on the problem. For example, the service functionality of an SI function block used to read an input could detect that the input module is without power supply and therefore detect that it could not perform the requested read of the input values depicted in the lower left service sequence diagram of Figure 5.1.

Apart from just describing the visible behaviour at the SI function block's interface, a further description about the service behaviour can also be given. Similar to the description of a function block's behaviour as described in Chapter 3 this can be done by adding service primitives on the resource's side of a service sequence diagram (e.g. `InitialiseService`, `ReceiveMessage`, `ReadInputs`). All service sequence diagrams in Figure 5.1 contain such descriptive service primitives.

Definition of service primitives

There are definitions for the meaning, i.e. semantics, of the positive and negative forms of all of the standard events supported by SI functions. IEC 61499 defines these various event forms as service primitives. Table 5.2 provides definitions for the standard service primitives. It has two columns that correspond to communications services and general purpose services respectively. Clearly, with complex SI function blocks all the various combinations and forms of normal and abnormal transactions utilising these service primitives should be considered.

Table 5.2 Definition of service primitives

Service primitive	Semantics when used in communications services	Semantics when used in general services
`INIT+`	Request to initialise communications service.	Request to initialise service.
`INIT−`	Request to terminate communications service.	Request to terminate service.
`INITO+`	Indication that communications service has been initialised.	Indication that service has been initialised.
`INITO−`	Indication that either a communications service could not be initialised or has been terminated successfully.	Indication that service has not been initialised or has terminated successfully.
`REQ+`	Normal request to transfer data.	Normal request for service.
`REQ−`	Disable request for data transfer.	Disable request for service.
`CNF+`	Confirmation of successful data transfer.	Normal confirmation of service.
`CNF−`	Confirmation that data transfer was unsuccessful.	Confirmation that service request was unsuccessful.
`IND+`	Indication of successful data arrival.	Indication of normal service arrival.
`IND−`	Indication of unsuccessful data arrival.	Indication of unsuccessful service arrival.
`RSP+`	Response by the application of successful data arrival.	Response by application of a successful service.
`RSP−`	Response by the application of unsuccessful data arrival.	Response by the application of an abnormal or unsuccessful service.

Partnered service interface function blocks

So far we have only considered SI function blocks which interact with a service provided by the resource. In several cases, there will be a paired set of SI function blocks which will interact with each other via services. For example, a pair of communication function blocks allows the exchange of data between different resources consisting of a sender and a receiver function block. Data given to the sender at its input is transferred to the receiver and provided at its outputs.

Similar to the service sequence diagrams to model the behaviour of adapter interfaces the standard specifies that service sequence diagrams may show the transactions for both partners. In this case, the resource part is omitted and the interfaces of both function blocks are used. Examples for the service sequence diagrams for this are shown in Figure 5.2. The two vertical bars represent the processing and time delays that occur to transfer a message between sender and receiver. Unlike the service sequence diagrams used for describing adapter interfaces for partnered SI function blocks, it is not necessary that both sides are involved in a service transaction. If only a local part of the service is invoked the other side will not be informed. This is shown in the `NormalInitialisation` service sequence diagram depicted in Figure 5.2.

The textual syntax for service sequence diagrams can include definitions of both ends of the transaction. For example, the type definitions for sender can include the expected behaviour of the receiver. To illustrate this, the following textual syntax would define the data transfer service sequence diagram as part of the sender's type definition, i.e. it describes the `NormalDataTransfer` service sequence diagram shown in Figure 5.2.

```
SEQUENCE NormalDataTransfer
   Sender.REQ+(SD_1,...,SD_m) -> Receiver.IND+(RD_1,...,RD_n);
  END_SEQUENCE
```

Note: The receiver's definition will simply refer to the sender for the full transaction definition. Also note that in this example the notation `SD_1, ...,` `SD_m` just indicates that the sender/receiver function blocks are general purpose and can be tailored to fit particular applications. In a real system, each

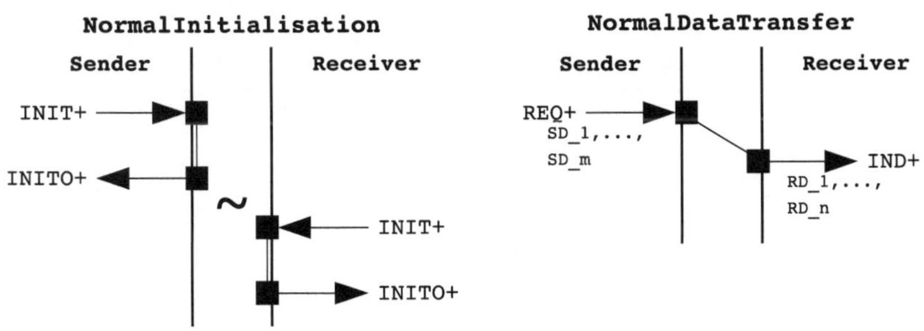

Figure 5.2 Service sequence diagrams for partnered service interface function blocks

sender/receiver pair would be specified to send and receive a fixed number of data items of a given data type.

A shortcoming from modelling the behaviour of partnered SI function blocks in this form is that the resource behaviour and the specific services used are not modelled. In order to provide this information additional service sequence diagrams describing only one of the both partnered SI function blocks need to be modelled.

5.3 Type definitions

IEC 61499 specifies in general terms the way in which an interface to an SI function block is defined. Every SI function block provides some kind of service, such as, reading I/O, publishing values over a network and so on. It is therefore proposed that the name given to each type of SI function block reflects the name of the service that the function block provides, e.g. `ValvePositioner, HMIWriter`.

5.3.1 General service interface function block types

It should be noted that an SI function block may initiate a service in response to a stimulus from an application. Alternatively, the SI function block may respond to an external request, such as, a signal from an HMI panel, to obtain a value from an application. SI function blocks can be modelled to support both types of behaviour.

Figure 5.3 depicts two SI function blocks that are given as examples and reference in the IEC 61499 standard. These function blocks are still 'generic' in the sense that they have not been given a specific number of input and outputs. Data types of inputs and outputs are also defined using the keyword `ANY`. In other words, these are outline templates for type definitions. Further specific details and data types will be required to use these in a particular application domain.

The `REQUESTER` function block provides a service for what IEC 61499 calls an 'application-initiated interaction' – in other words; the request for external data is triggered by an event generated within an application. The `PARAMS` input is used to specify the service specific configuration parameters; inputs `SD_1, ... SD_m` provide parameters for a request, e.g. by sending a request command string and

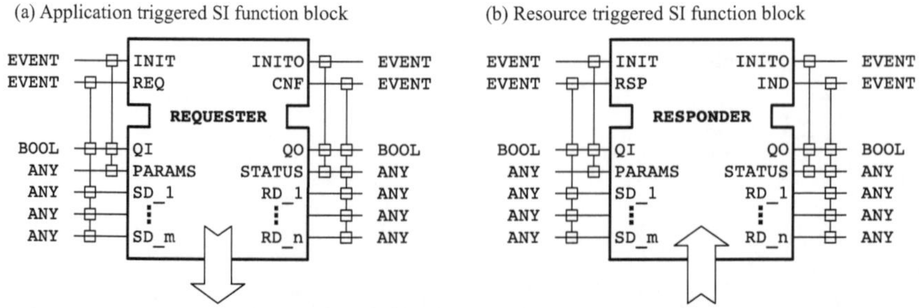

Figure 5.3 Application and resource triggered service interface function block

arguments. The RD_1, ... RD_n outputs are used for the response data received from the service and will be provided when the function block issues a confirmation CNF output event.

In contrast, the RESPONDER function block provides a service for a 'resource-initiated interaction'. The request for data arrives from an external resource and results in an indication IND event together with output data on RD_1, ... RD_n being produced. In this case, the application will need to react to an event that can occur at any time. The response to the request can be returned, by writing the data to the RESPONDER inputs SD_1, ... SD_m and triggering an event at the response RSP event input notifying the service that the processing of the received data has been completed.

This behaviour of the two general SI function block types is shown in the form of service sequence diagrams in Figure 5.4. These diagrams show the typical life-cycle for the function blocks consisting of initialisation, normal usage and finally termination. While the first and last service sequence in the service sequence diagrams for REQUESTER and RESPONDER are the same the main difference is in the NormalDataTransfer service sequence of both types. Here it is clearly visible that the REQUESTER will perform its service request on the reception of an REQ

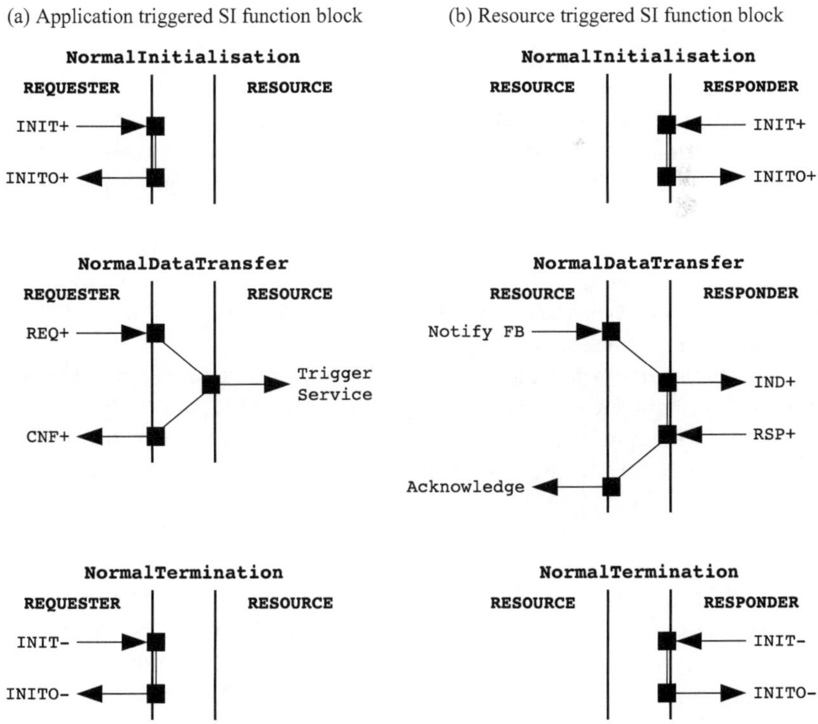

Figure 5.4 Service sequence diagram for application and resource triggered interactions

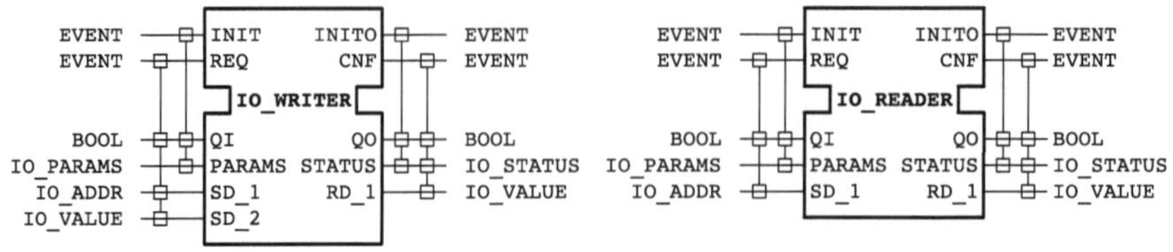

Figure 5.5 IO_WRITER and IO_READER service interface function block

event from the application while in the RESPONDER case the service takes over the active part and informs the application via an IND event.

Together, the REQUESTER and RESPONDER SI function blocks can be used to exchange data between two resources linked by some form of communications or networking facility. Therefore typically a set of partnered SI function blocks, as described in the last section, consist of a REQUESTER and RESPONDER SI function blocks.

5.3.2 Example service interface function block types

Figure 5.5 shows two further examples of SI function blocks; IO_WRITER is used to write values to physical outputs while IO_READER is provided to read in values from selected physical inputs. Both function in a similar manner.

Let us consider how the IO_WRITER function block could be used. An application will require at least one instance of this function block to write values out to physical outputs. To set-up the service, the application must first send an INIT event with QI input set to TRUE and the PARAMS input set to identify characteristics of the service. The PARAMS input could contain details, such as 'write' update rates, number of retries on failure and so on. Thereafter, data can be sent to the selected output by setting the input SD_1 to an output address, e.g. rack, channel and I/O point and setting the new value to input SD_2. The write is initiated by an event at event input REQ.

Some time later when the hardware I/O system has completed the write operation, an output event CNF will occur to confirm that the 'write' has been completed. The output STATUS provides an indication of whether the operation has been successful. If the operation has failed, STATUS will contain an appropriate error code. The output RD_1 provides a feedback of the value as read from the output device. This can be used to confirm that the write has been successful.

The IO_READER function block performs in a very similar manner. After initialising the service using the INIT event, with the QI and PARAMS inputs, the value of any physical input can be read by setting the IO address at input SD_1 and sending an event to event input REQ. Some time later, when the data has been read from the input sensor, a confirmation event will occur at output event CNF. The success of the read operation will be indicated by the value of output STATUS and if successful, the value read from the input will be available on output RD_1.

Note: In this example for both function blocks the application triggered SI function block type has been applied. For the `IO_READER` the resource triggered scheme could also be an option. In this case the `PARAM` input would also contain information on the IO address and on `INIT` the function block would register this IO address. The service can then notify the application with an `IND` event when a new value is available or a certain time period has elapsed.

`IO_WRITER` and `IO_READER` are fairly simple forms of SI function blocks for accessing hardware I/O. However, more complex function blocks could be considered to model facilities to read and write, say, multiple values for many I/O points in one operation.

Note that the time taken between issuing a request by triggering an `REQ` event and receiving the confirmation event `CNF` will depend on many factors such as:

- the loading of the resource scheduling system,
- the speed of the device operating system in responding to the request from the resource,
- the time to transmit a request to the physical I/O point.

5.3.3 Textual syntax – service interface function block example

The textual syntax for defining SI function blocks can be best demonstrated by reviewing the textual type definition for the `IO_WRITER` function block as shown in Figure 5.5.

```
FUNCTION_BLOCK IO_WRITER (* IO_Writer Service Interface *)
  (* Event definitions *)
  EVENT_INPUT
    INIT WITH QI, PARAMS;
    REQ WITH QI, SD_1, SD_2;
  END_EVENT
  EVENT_OUTPUT
    INITO WITH QO, STATUS;
    CNF WITH QO, STATUS, RD_1;
  END_EVENT
  (* Variable definitions *)
  VAR_INPUT
    QI : BOOL;             (* Event input qualifier *)
    PARAMS : IO_PARAMS;    (* Service parameters *)
    SD_1 : IO_ADDR;        (* Output address *)
    SD_2 : IO_VALUE;       (* Output value *)
  END_VAR
  VAR_OUTPUT
    QO : BOOL;             (* Event output qualifier *)
    STATUS : ANY;          (* Service status *)
    RD_1 : IO_VALUE        (* Returned value *)
  END_VAR
```

```
  (* Service sequence definitions *)
  SERVICE IO_WRITER/RESOURCE
    SEQUENCE normal_initialisation
      REQUESTER.INIT+(PARAMS) -> REQUESTER.INITO+(STATUS);
    END_SEQUENCE
    SEQUENCE abnormal_initialisation
      REQUESTER.INIT+(PARAMS) -> REQUESTER.INITO-(STATUS);
    END_SEQUENCE
    SEQUENCE normal_data_transfer
      REQUESTER.REQ+(SD_1,SD_2) -> REQUESTER.CNF+(RD_1);
    END_SEQUENCE
    SEQUENCE abnormal_data_transfer
      REQUESTER.REQ+(SD_1,SD_2) -> REQUESTER.CNF-(STATUS);
    END_SEQUENCE
    SEQUENCE normal_termination
      REQUESTER.INIT-() -> REQUESTER.INITO-();
    END_SEQUENCE
    SEQUENCE abnormal_termination
      REQUESTER.INIT-() -> REQUESTER.INITO-(STATUS);
    END_SEQUENCE
    SEQUENCE resource_initiated_termination
      -> REQUESTER.INITO-(STATUS)
    END_SEQUENCE
  END_SERVICE
END_FUNCTION_BLOCK
```

Comparing this with the general interface specification from Chapter 3, we can see that an SI function block does not contain any additional information to what was already described there. Therefore, the type definition for an SI function block provides the minimal information needed to use a function block as part of applications, subapplications or composite function blocks.

Note: In the last sequence definition for `resource_initiated_termination` there is no component to the transaction given to the left of the '−>' symbol. This is because, in this case the transaction is produced by some unknown agent outside the resource. For example, this might occur when there is an imminent power-supply failure and the device is therefore sending an unsolicited transaction across the resource interface to terminate the `IO_WRITER` service.

Any additional behaviour of the SI function block, for example, to check the validity of particular inputs prior to initiating a service, can be modelled by encapsulating the SI function block within a composite function block as discussed in Chapter 3. Extra function blocks could then be added to filter input data and verify response data.

It should be noted that the SI function block type definition only contains definitions of behaviour that occurs within the function block domain, i.e. within the resource. It does not define the behaviour outside the resource that is provided by the device operating system, hardware or communications systems. This is all outside the scope of IEC 61499.

5.4 Communication service interface function blocks

As IEC 61499 targets distributed industrial process measurement and control systems, communication between the application parts located in different resources is an important issue. So far we have only considered communication SI function blocks as a form of SI function block that provides communication services to the application utilising device and resource functionalities; for example, in Chapter 2 when discussing the device specific configurations and adaptations of mapped applications.

Two different communication models with associated generic SI function blocks are defined in IEC 61499 serving the main needs of distributed applications. These two models are the unidirectional transaction and the bidirectional transaction models.

5.4.1 Function blocks for unidirectional transactions

The unidirectional transaction model can be utilised to send data from a resource to one or more other resources. For this model IEC 61499 defines the `PUBLISH/SUBSCRIBE` function block pair as shown in Figure 5.6. With these function blocks an application can send data with the `PUBLISH` function block to be received by one or more `SUBSCRIBE` function blocks. Hereby the sender is completely decoupled from the receiving entities. That means `SUBSCRIBE` function blocks can be added or removed without influencing other `SUBSCRIBE` function blocks or the `PUBLISH` function block. As this corresponds to the functionality of data or event connections in applications the unidirectional transaction is typically used for supporting data and event connections of distributed applications across resource boundaries as described in Section 2.4.

The main service sequence diagrams describing this unidirectional transaction are shown in Figure 5.7. On initialisation with an `INIT` input event both function blocks connect to their local communication service (e.g. network driver) and

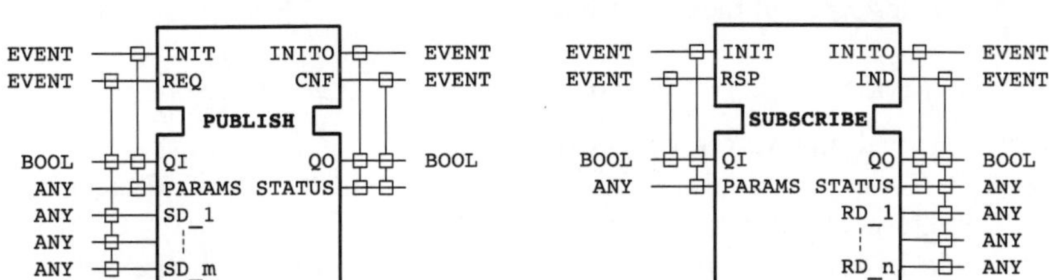

Figure 5.6 PUBLISH and SUBSCRIBE communication service interface function blocks

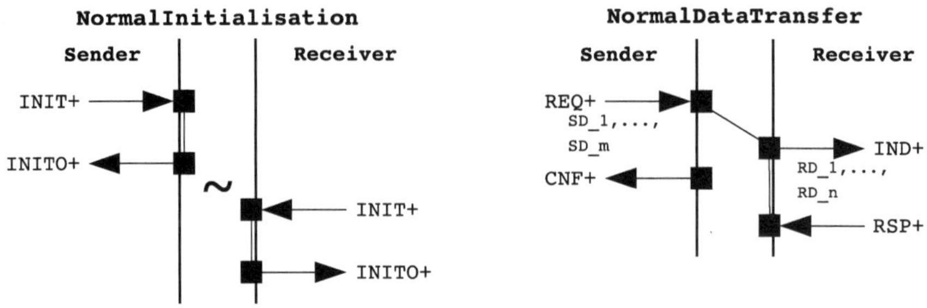

Figure 5.7 Main service sequence diagrams describing the unidirectional transaction

perform all initialisations required for sending or receiving data. Both sides can perform their initialisation independent from each other. The parameters given in the PARAMS input specify how the PUBLISH function block will produce messages which can then be received, identified and decoded by the corresponding SUBSCRIBE function block (e.g. through common message identifiers and common message encoding). Similar to the other presented SI function blocks, the values of the PARAMS input are implementation and network specific.

During normal operation a transaction is started by the triggering of the PUBLISH function block with an REQ input event. The PUBLISH function block takes the input data (i.e. SD_1, ... SD_m) and generates the appropriate message to be transmitted on the network. After successfully sending the message a CNF output event is issued at the PUBLISH function block. On the reception of a message at the SUBSCRIBE function block, the message's data is applied to the outputs RD_1, ... RD_n and an IND event is issued. The application notifies the SUBSCRIBE function block on the reception of the data with an RSP event.

Note: Triggering of the PUBLISH's CNF output event is neither linked with the successful reception of the sent message at the SUBSCRIBE function block nor with the RSP event at the SUBSCRIBE function block. Therefore, a PUBLISH function block has no information of whether the data sent has been received or processed successfully. Only the information on the status of the successful send data transmission is available. If an application requires status information on the delivery of the messages sent, other means have to be utilised (e.g. the bidirectional transactions described in the next section).

5.4.2 Function blocks for bidirectional transactions

The bidirectional transaction model can be used when an application is distributed between two resources A and B, and there is a requirement to send request data from A and receive a response from B. IEC 61499 provides CLIENT/SERVER SI function blocks for this form of interaction that are capable of bidirectional data transfer (Figure 5.8). The CLIENT function block in resource A sends out the request to the SERVER function block in resource B, the SERVER function block responds and sends the response data back to the CLIENT function block in A.

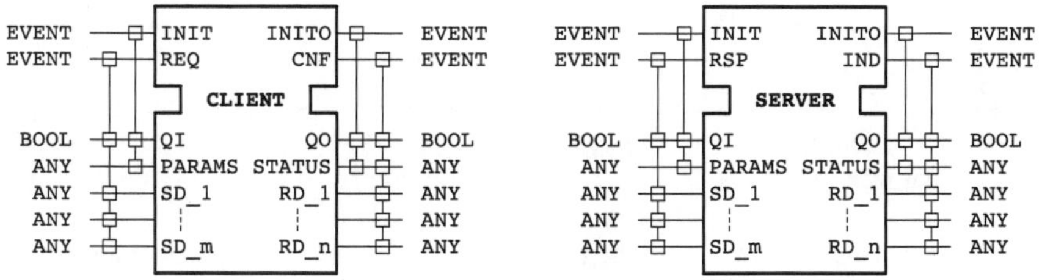

Figure 5.8 CLIENT and SERVER communication service interface function blocks

The service sequence diagrams for the bidirectional transaction as depicted in Figure 5.9 clearly show the significant differences to the unidirectional transaction. The first main difference is in the initialisation sequence. It is necessary that the SERVER function block is initialised before the associated CLIENT function block can perform its initialisation. The initialisation of the SERVER function block is equivalent to that in the unidirectional transaction model. On the INIT event the SERVER function block initialises its local services for receiving and sending messages. When the CLIENT function block performs its initialisation it sends a registering message to the SERVER function block (e.g. establish a connection). In the service sequence diagram this is indicated with the NULL service primitive.

Similar to the PUBLISH function block, the CLIENT function block will take its input data (i.e. SD_1, ... SD_m) to create a network message and send it on the reception of an REQ event. However, unlike the unidirectional model, the CLIENT function block will wait for a response from the SERVER function block. On the reception of a message, the SERVER function block will update its outputs (i.e. RD_1, ... RD_n) accordingly and notify the application by issuing an IND event. The application processes the data and generates an appropriate response by setting the SERVER function block's input data and triggering the RSP event. The SERVER function block puts this data into a network message and sends it back to the CLIENT function block. There the received data is made available at the

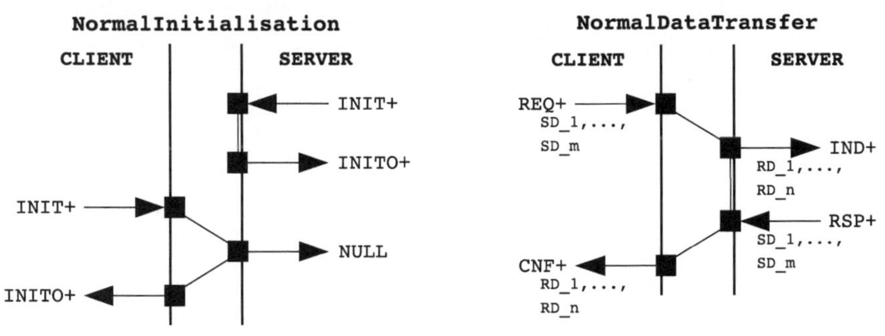

Figure 5.9 Main service sequence diagrams describing the bidirectional transaction

CLIENT function block's outputs and the application is informed by issuing the CNF event.

A CLIENT and SERVER pair in effect provides a mechanism to create a local 'proxy' of a remote function block. They can provide a local set of inputs and outputs that mimic those available on the remote function block. It is envisaged that engineering support tools will be able to automatically insert CLIENT/SERVER or PUBLISH/SUBSCRIBE pairs when an application is split between resources.

5.5 Management function blocks

The last form of SI function block that is defined in the standard concerns services to load, start and initiate function block execution. The standard defines a generic form of management function block as shown in Figure 5.10 that can be used to initiate a range of service functions using different command definitions. The standard defines nine commands (listed in Tables 5.3 and 5.4) which provide the means necessary for the configuration and reconfiguration tasks required for managing distributed applications. The descriptions of these services and how they work is only given in outline in the standard. This is an area which is particularly difficult to model because systems will tend to have completely different mechanisms for loading and creating function block networks and starting the execution of loaded applications. This is a large topic that will be difficult to model in a general way. Therefore the IEC 61499 standard working group re-assigned this task so that it is included in the development of compliance profiles under the topic 'configurability provisions' (see Section 8.3 and Appendix B).

At one extreme, to start an application, a system may require that all function blocks and supporting resource libraries are compiled to a binary format and downloaded into separate devices. At the other extreme, devices may have large libraries of pre-loaded function blocks; for example, function block definitions could all be part of the device firmware. In this case, an application can be created by simply downloading the definitions of the connections that are required to create the application's function block networks in the different devices. There may also be hybrid systems where some function blocks are in firmware while others are downloaded. The services that are supported by the management function block apply at both the resource and device level as shown in Tables 5.3 and 5.4

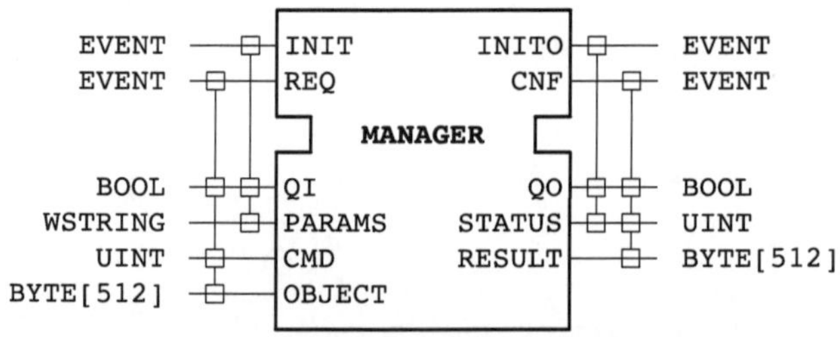

Figure 5.10 Management function block

Table 5.3 Resource level management service functions

Service function	Description
Create	The Create service establishes the data type definitions, function block types and instances, and connections between function blocks. This will involve downloading definitions from a source, e.g. copying across a network, copying in from a memory card.
Write	The Write service initiates the configuration of parameters of function block instances.
Start	The Start service triggers the execution of function block networks within a resource. Typically it will start the resource scheduling function, put function blocks into the Running state (i.e. enable input event reception, see Section 2.5.2) and start to run SI function blocks that generate timing events. These in turn trigger chains of events that cause function block execution.
Stop	The Stop service causes all execution to cease by ending ECC evaluation, stopping event input delivery, but finishing currently pending algorithms.
Kill	The Kill service causes all execution to cease immediately by interrupting currently executed algorithms and suspending the resource scheduling function.
Reset	The Reset service puts the function block back into its initial sate. This concerns setting up function blocks and will include resetting variables to their default initial values.
Delete	The Delete service can be used to delete the definition of any data type, function block or connection.
Query	The Query service provides a means to access information about the status, existence of data types, function blocks types, function block instances and their connections.
Read	The Read service provides a means to access the current value of function block instance parameters.

Table 5.4 Device level management service functions

Service function	Description
Create	Creates and establish a resource within a device.
Write	This sets up the attributes and parameters of a resource in a device.
Start	Starts the resource to allow function block execution.
Stop	Stops the resource execution of function blocks.
Kill	Immediately stops the resource execution of function blocks.
Reset	Resets the resource into the initial state after creation enabling it to be started again.
Delete	Deletes a resource from the device so that it can no longer be accessed to load and execute function blocks.
Query	Queries the device to obtain information regarding the status and existence of resource(s) in the device.
Read	Reads the values of resource parameters.

The Manager function block can be used by specifying different values for the CMD input to initiate the various service functions. The response to each form of service function is given by the value returned in the `RESULT` output. The service function is further characterised by the value of input `OBJECT`.

5.5.1 Examples

1. Create a function block instance in a resource:

    `CMD` = **CREATE**
    `OBJECT` = *function block instance definition data*
    `PARAMS` = resource *reference*
    `RESULT` = *function block instance reference*

2. Utilise Query management service function to find the reference to the end-connection for a given start-connection:

    `CMD` = **QUERY**
    `OBJECT` = *connection start point definition*
    `PARAMS` = *resource reference*
    `RESULT` = *connection end point definition*

Note: In these two examples the data types for values shown in italics are not defined in IEC 61499 and are regarded as implementation specific. However, the standard has defined a data exchange format for porting function block definitions based on XML, see Section 2.6.

As a general comment, function blocks are very effective at modelling systems where the behaviour is concerned primarily with data and event flows. It is questionable whether management operations for loading and creating function blocks to form applications that are distributed across devices can be best described using the same model. This type of behaviour may require a different type of model, as it is mainly concerned with data management issues.

5.5.2 Managed function blocks

Function blocks that can be loaded and accessed by management function blocks are called 'managed function blocks'. This is in contrast to non-managed function blocks; these would include function blocks that are part of device firmware and therefore have a certain degree of fixed functionality.

Managed function blocks can have a number of states as already discussed in Section 2.5. The state of a managed function block can be changed if it is associated with a Manager function block. The concept is that a resource can dynamically change the configuration of function blocks by issuing requests to Manager function blocks. The way this will actually occur is rather sketchy in the standard. However, use of these concepts in applications have shown how this can be utilised for performing dynamic reconfiguration of control applications during full operation of the plant [20].

5.6 Summary

We have now covered the important features of IEC 61499 Service Interface function blocks. To re-cap, service interface function blocks are provided to give an interface between function blocks running in the resource and services provided by or outside the resource. A service interface function block type definition only defines the interface into the service and its response; it does not define the implementation of the service behaviour. Service sequence diagrams are used to show and describe the timing relationships between events on the input and output side of service interface function blocks.

Service interface function blocks can be used to:

- read and write to I/O points accessed via the device's I/O sub-system,
- request data from external resources or respond to data requests from external resources,
- maintain event and data connections across resource boundaries, e.g. using `PUBLISH/SUBSCRIBE` function blocks,
- set-up interlocked exchanges of data with external resources, e.g. using `CLIENT/SERVER` function blocks,
- manage the creation and execution of function blocks in a resource.

Chapter 6
Event function blocks

As discussed in Chapters 1 and 2, a fundamental property of IEC 61499 function block networks is that the execution of all software within every function block is, in some way, event triggered. In this chapter, we will review a special set of standard function blocks that are provided for event behaviour. These function blocks can be used to model the control, generation and detection of events.

This chapter describes the standard event function blocks that:

- *allow events to be split to produce new events,*
- *allow events to be merged,*
- *permit the propagation of particular events,*
- *select between two or more events,*
- *delay an event by a given period,*
- *generate streams of events,*
- *create events from Boolean edge detection,*
- *count the number of event occurrences.*

6.1 Overview

In the previous chapters, we have reviewed several different ways to create function block type definitions, i.e. for basic, composite and service interface function blocks. In this chapter, we will review a special set of function block types that are defined in IEC 61499 specifically to handle events. Most of these function blocks are defined as basic function blocks. A couple of blocks are defined as a form of service interface function blocks as they require some interaction with the containing resource. There are also some more complex event function blocks which are built as composite function blocks from the elementary event function blocks.

All the standard event function blocks are designed to be used within the definitions of larger user composite function blocks and applications and provide many of the common operations required when modelling event behaviours.

In any event oriented system there is always a need to be able to generate the initial event that triggers the execution of the first item of software. Likewise, in IEC 61499 function block networks, an event is required to trigger the first function block in every chain of function blocks. Control systems will typically also have other requirements such as a need to split events to simultaneously trigger multiple chains of function blocks or to create trains of events so that function blocks are regularly re-executed, e.g. when modelling a system that scans I/O.

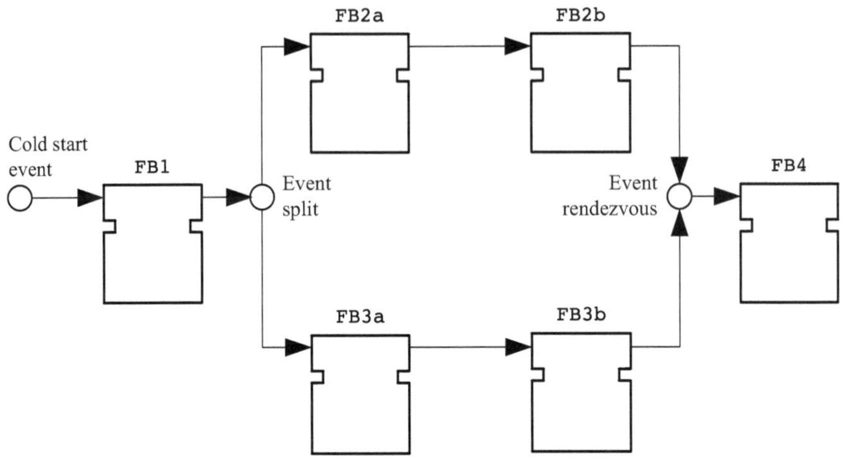

Figure 6.1 Event chain example

Figure 6.1 shows a simple example of a control system that starts execution when the system is initially powered-up, i.e. 'cold started' and then goes on to execute two chains of function blocks in parallel. A 'cold start' event is clearly needed to initiate the first function block in the chain. When the first function block FB1 has completed its execution, there is a requirement to execute two separate chains of function blocks. This is shown by the event splitting and triggering function blocks FB2a and FB3a. When FB2b and FB3b have both completed their execution, events from the two function blocks are joined using a 'rendezvous' to create a single event that then in turn fires the last function block FB4.

In this first example, the function blocks only execute once when the system cold starts, because only one initial event is generated. However, typically a control system will require scanned behaviour where function blocks need to execute repeatedly to monitor plant states and update actuators and so on. This can be achieved by passing a train of events into the first function block, as shown in Figure 6.2. Assume that the event train consists of a regular stream of events sent at regular intervals, say every, 100 ms. Now all function blocks in both parallel paths will be continually re-executed every 100 ms after cold start.

In Figures 6.1 and 6.2, each circle represents some form of event propagation behaviour. In fact, we will see that each circle can be replaced by one of the IEC 61499 event function blocks. The timing diagram shown below the function block network in Figure 6.2 depicts the stream of events entering the event control header of function block FB1. The event stream emerging from the event rendezvous is also shown. Notice that here the events are slightly delayed because of the time to execute algorithms in the parallel function block networks and because of delays in resource scheduling. The period between events also varies slightly because execution delays may vary, for example, an algorithm may not necessarily always take the same period of time to execute as this may depend on the values of input and internal variables.

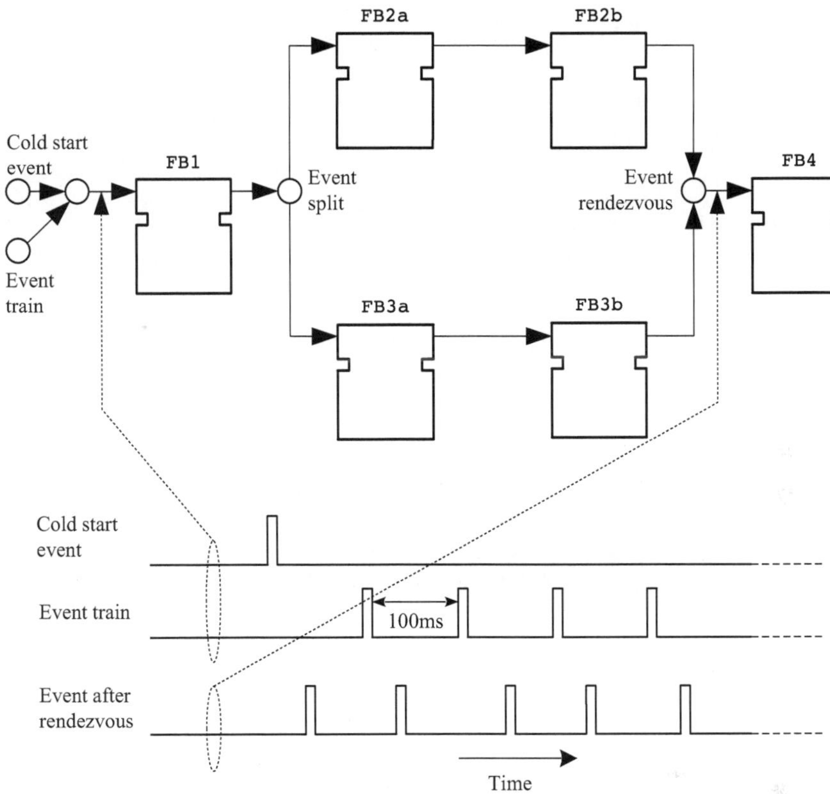

Figure 6.2 Cyclic event chain example

6.2 Standard event function block types

We will now review the rich repertoire of standard event function block types defined in IEC 61499 that allows a wide range of different event scenarios to be modelled. In the following descriptions each figure shows the interface of the function block type on the left of the figure and its corresponding type specification on the right. Many of these event function blocks are basic function blocks and can be defined using an Execution Control Chart (ECC); some require interactions with the resource and are therefore defined using service sequence diagrams. There are a few complex event function blocks, such as the 'Boolean Rising Edge Detector', that are defined as composite function blocks formed from simpler event function blocks.

Note: In the function block type declarations that are shown in the following figures, the formal IEC 61499 function block type name is shown inside the function block figure. Where there are associated algorithms required, these are depicted below the ECC in the function block type specification. In some cases, timing diagrams are also depicted below some of the function blocks to clarify the timing relationships between input and output events.

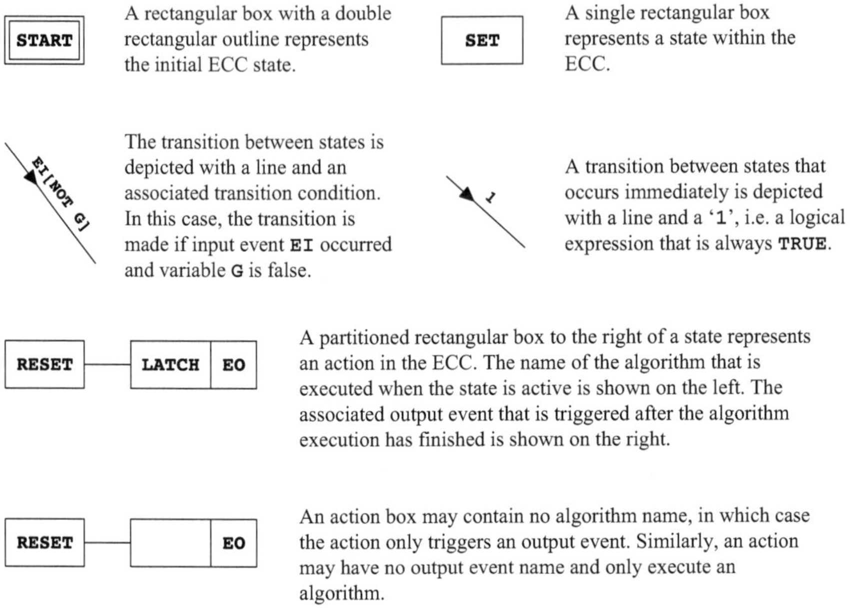

Figure 6.3 Understanding the ECC notation

It is surprising to find that for many of these function blocks the internal behaviour can be completely described by just using an ECC. As an aid to understanding the following ECCs, the points shown in Figure 6.3 summarise the main aspects of the ECC notation that is fully described in Section 3.4.

In some of the following definitions such as the Event Merger, the number of inputs or outputs may vary. For example, the Event Merger function block may merge two or more events. In such cases, it is envisaged that software tools will automatically dimension function block instances to have the required number of inputs and outputs.

6.2.1 Event Splitter

The Event Splitter produces two or more simultaneous output events on the reception of a single input event, as shown in the timing diagram (Figure 6.4).

6.2.2 Event Merger

The Event Merger produces a single output event stream by merging the incoming events on the input events `EI1` to `EIn`. This can be used to merge two or more events (Figure 6.5).

6.2.3 Two Event Rendezvous

The Two Event Rendezvous waits for the reception of one event on both input events `EI1` and `EI2` and produces a single output event on output `EO`. If the function block receives a further single event on both `EI1` and `EI2` inputs, it will produce a further output event on `EO` (Figure 6.6).

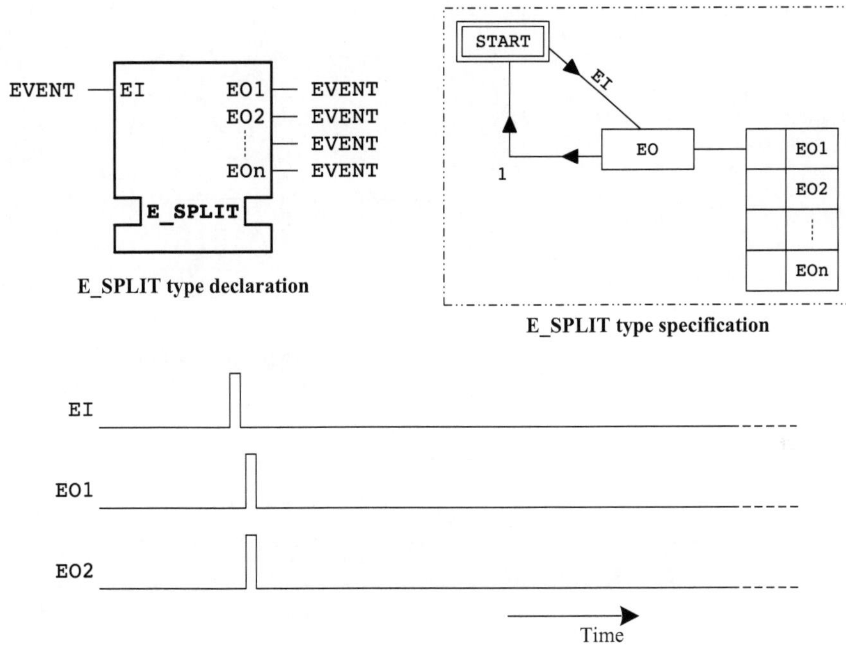

Figure 6.4 Event Splitter

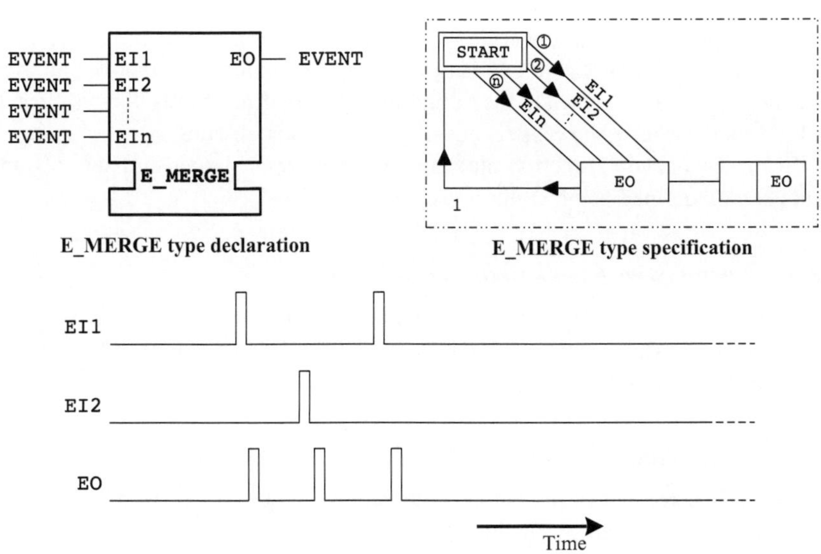

Figure 6.5 Event Merger

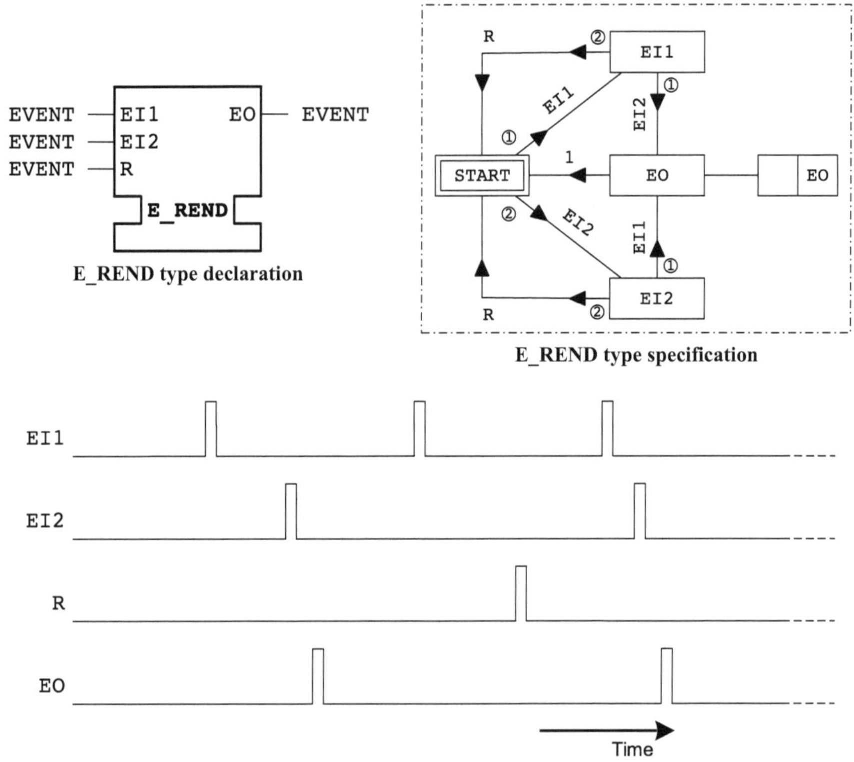

Figure 6.6 Two Event Rendezvous

Note if at any time the function block receives a reset event on input R then a single event on both EI1 and EI2 is always required before a further output event on EO is produced, as shown in the timing diagram. In other words, the reset can be used to re-synchronise the rendezvous of the two event streams.

Event Rendezvous function blocks that handle other larger numbers of input events can be defined using similar designs.

6.2.4 Permissive Event Propagator

The Permissive Event Propagator controls the flow of events from input EI to EO depending on the value of input PERMIT. A TRUE value is required to allow events to flow through the function block (Figure 6.7).

6.2.5 Event Selector

The Event Selector allows the output event stream to be selected from either event input EI0 or EI1 depending on whether G has the value FALSE or TRUE, respectively (Figure 6.8).

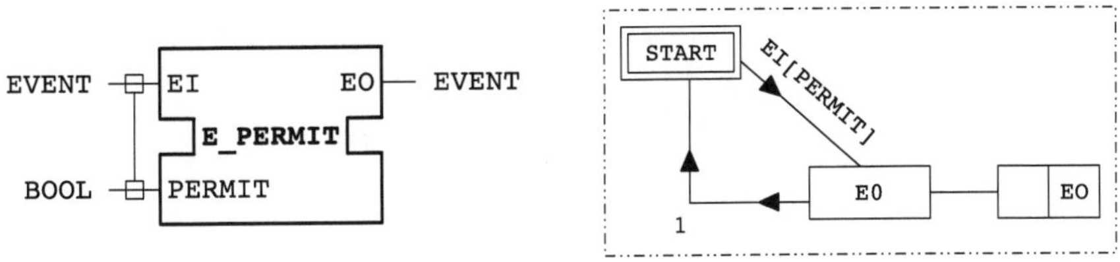

E_PERMIT type declaration

E_PERMIT type specification

Figure 6.7 Permissive Event Propagator

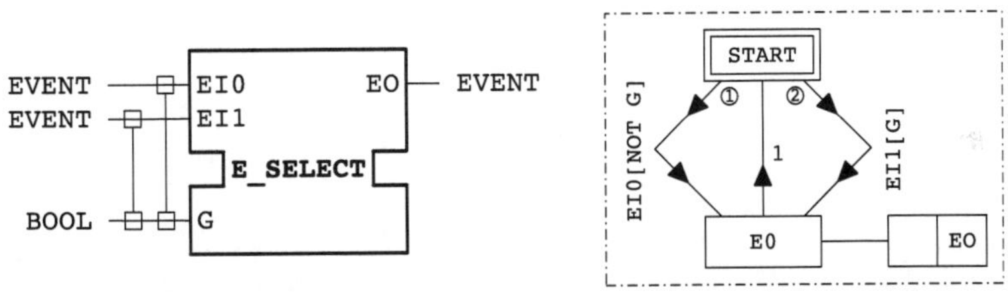

E_SELECT type declaration

E_SELECT type specification

Figure 6.8 Event Selector

6.2.6 Event Switch

The Event Switch can be used to steer the incoming event steam on `EI` to go to the output event `EO0` or `EO1`. `G` set to `FALSE` switches the events to output `EO0`; `G` set to `TRUE` switches the events to output `EO1` (Figure 6.9).

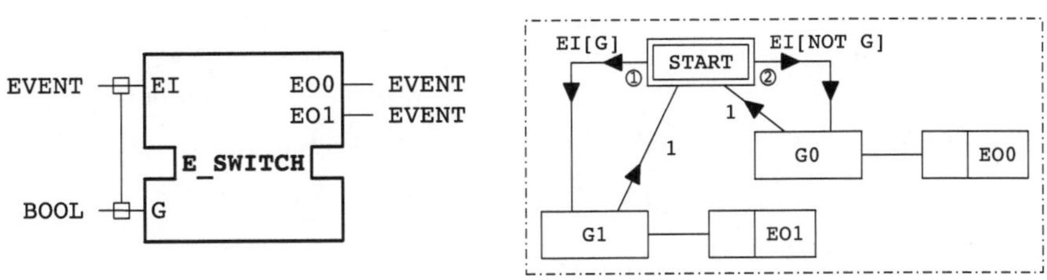

E_SWITCH type declaration

E_SWITCH type specification

Figure 6.9 Event Switch

6.2.7 Delayed Event Propagator

The Delayed Event Propagator provides a means to delay an event arriving at input `START` by a defined period as given by input `DT`. For example, if `DT` is set to 100 ms, then an event at output `EO` will be produced 100 ms after every event arriving at input `START`, provided the input events are at least 100 ms apart (Figure 6.10).

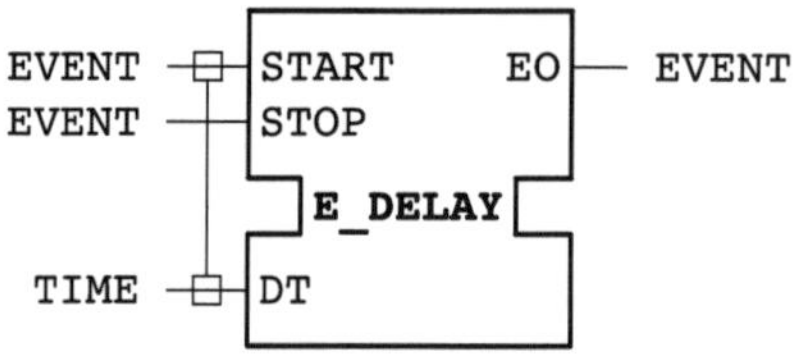

E_DELAY type declaration

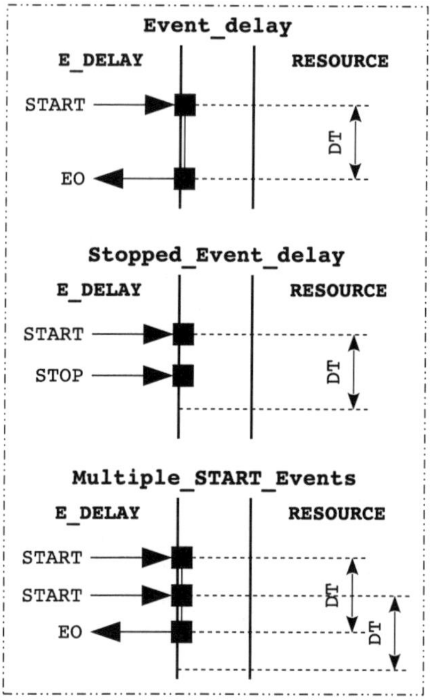

E_DELAY type time-sequence diagram

Figure 6.10 Delayed Event Propagator

However, if further events arrive at input START before the output event at EO has been produced, these will be ignored because there is no input event queue. If a STOP event is received while an input event is being delayed, it will be cancelled and no output event will be produced.

Note: This function block requires interactions with the underlying resource and therefore it is described using a simple set of service sequence diagrams. Generally, timing can only be achieved by some form of interaction outside of the resource, for example, through use of a timer circuit in the device sub-system.

6.2.8 Restart Event Generator

The Restart Event Generator provides single events when the resource cold and warm starts and also when the resource is stopped by some external agent. It is likely that most applications will require at least one instance of this function block to initiate the triggering event that starts execution of a chain of function blocks in a network, as shown in the Figures 6.1 and 6.2. This function block requires interactions with the resource and is therefore described using a service sequence diagram (Figure 6.11).

6.2.9 Cyclic Event Generator

The Cyclic Event Generator is used to produce a stream of events at regular intervals. The event stream starts when an event arrives at input START and stops

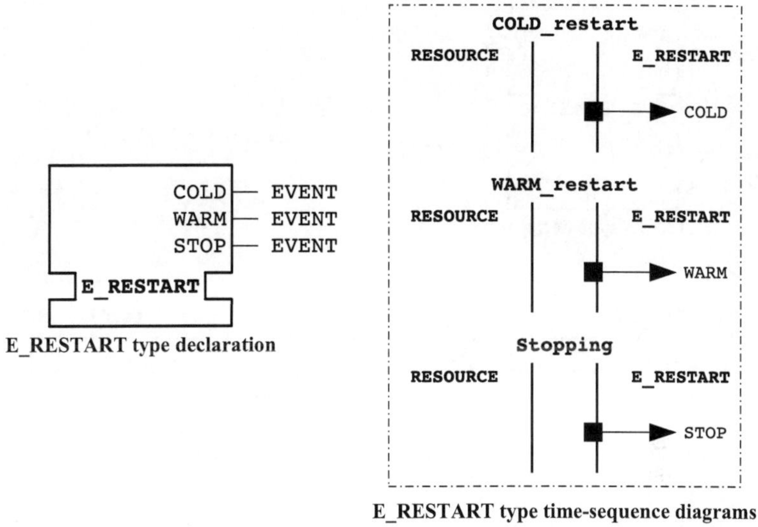

E_RESTART type declaration

E_RESTART type time-sequence diagrams

Figure 6.11 Restart Event Generator

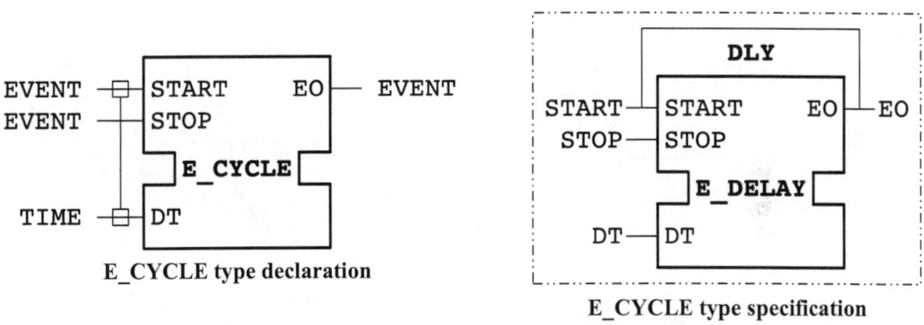

E_CYCLE type declaration

E_CYCLE type specification

Figure 6.12 Cyclic Event Generator

when an event arrives at input STOP. The period between output events is set by the value of DT.

This function block is described as a composite function block using an instance of Delayed Event Propagator E_DELAY. The feedback path from EO to START ensures that a new event is delayed after the initial start event, and subsequently after every output event (Figure 6.12).

6.2.10 Event Train Generator

The Event Train Generator produces a train of events at output EO each separated by a given interval. The number of events is set by the value of input N and the interval is defined by the value of input DT. The count of the number of output events that have been produced is given on output CV.

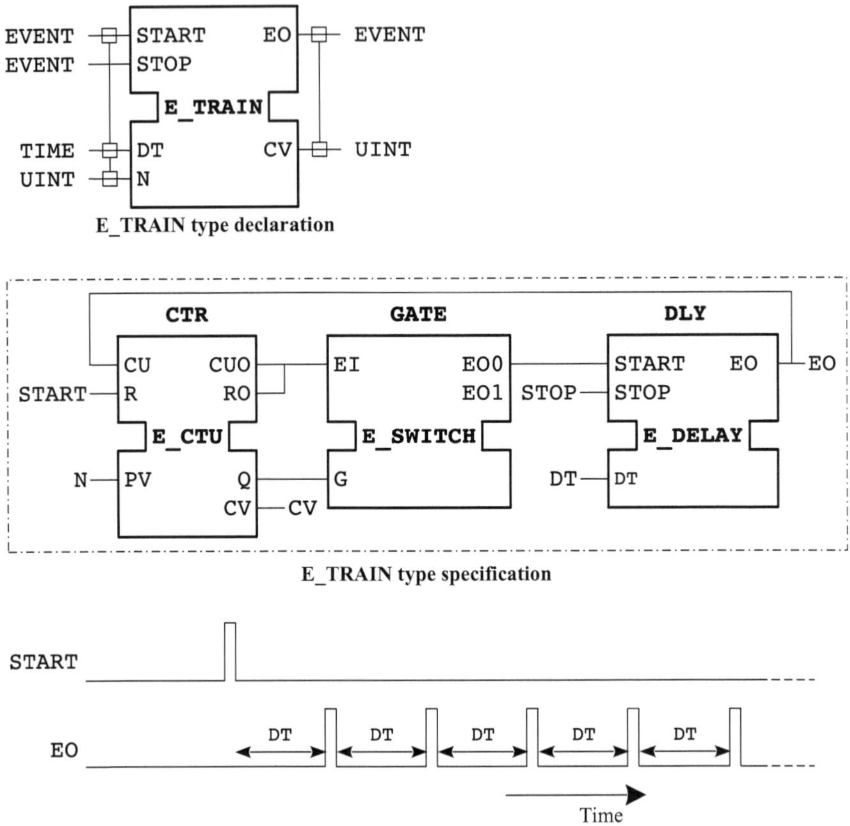

Figure 6.13 Event Train Generator

If an input event is received on input STOP, the event output train will cease. A further START event will cause a new train of output events to be produced (Figure 6.13).

This function block is defined as a composite function block using instances of the Event-driven Up Counter, the Event Switch and Delayed Event Propagator function blocks; these are all described in this chapter.

6.2.11 Event Train Table-driven Generator

The Event Train Table-driven Generator provides a train of events on output EO where the number of events and the intervals between events can be specified. The number of events is given by input N, the TIME array given at input DT defines the interval between each sequential output event. For example, if DT is set to the value of a time array of size 4, i.e. TIME[4], then the interval between the first and second events will be the value of TIME[0], between the second and third the value of TIME[1] and so on (Figure 6.14).

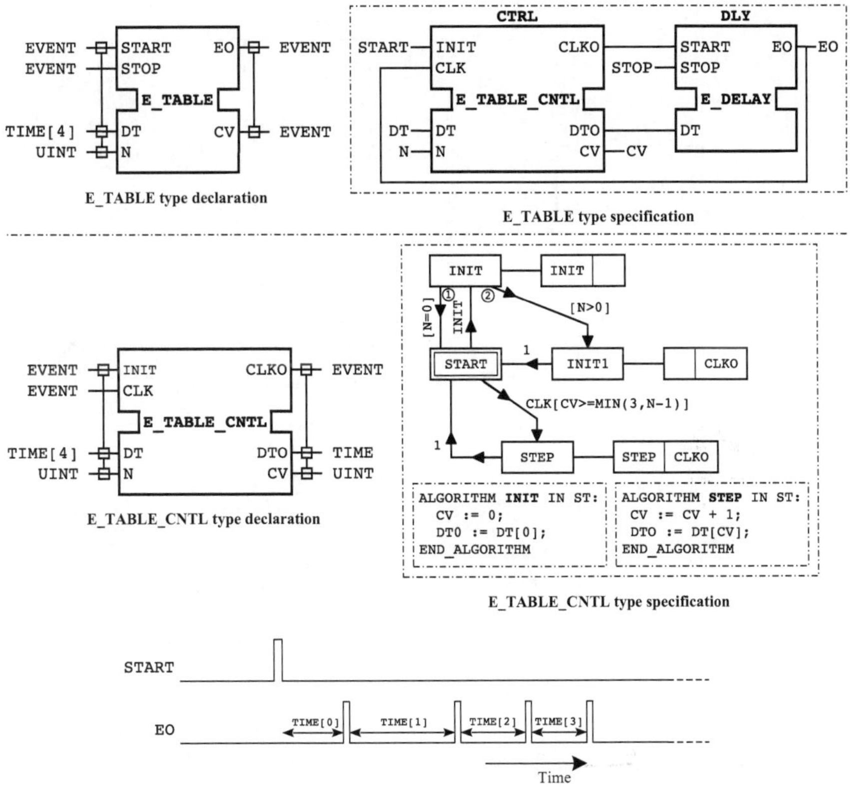

Figure 6.14 Event Train Table-driven Generator

Note: If the size of the array is m, the value of N should be m+1. For example, if there are five output events in the train, then four intervals are required in the time array.

This function block is refined as composite function block, using instances of the Table Control function block E_TABLE_CNTL and Delayed Event Propagator E_DELAY. The E_TABLE_CNTL is a special support function block used to select delay times from the TIME[4] array as the output events are counted. In this case, it is designed for an array of size 4 but could be used for arrays of any size by changing the constants in the expressions, where CV is tested.

6.2.12 Separate Event Table-driven Generator

The Separate Event Table-driven Generator is used to produce single events at each of the event outputs; each event separated by a defined interval. The number of events is defined by input N and the time interval between output events is given by the time array set on input DT (Figure 6.15).

For example, assume that the input to DT is a time array of size 4, i.e. TIME [4], and input N is set to 4 and the function block is defined to have 4 output

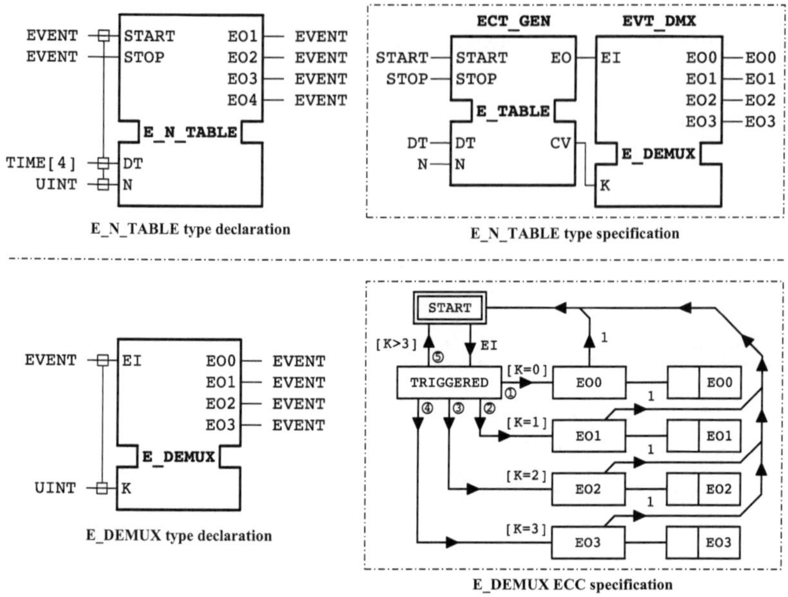

Figure 6.15 Separate Event Table-driven Generator

events. Then when an event arrives at input START, an output event at EO0 will be produced after a period TIME[0], a further output event will be produced at output event EO1 after a period TIME[1] and so on until all 4 output events have fired once. If an event is received on input STOP, the generation of output events ceases. A subsequent START event can then trigger a new set of output events.

This function block is defined as composite function block, using instances of the Event Train Table-driven Generator and an Event Demultiplexor.

6.2.13 Event-driven Bistable

The Event-driven Bistable provides a means to latch the arrival of an event. Its output Q is set to TRUE when an event arrives at the 'set' input S. The output Q is cleared to FALSE when an event arrives at the 'reset' input R (Figure 6.16).

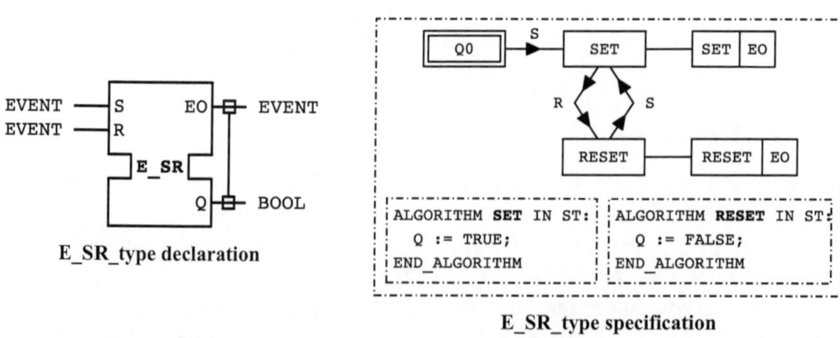

Figure 6.16 Event-driven Bistable

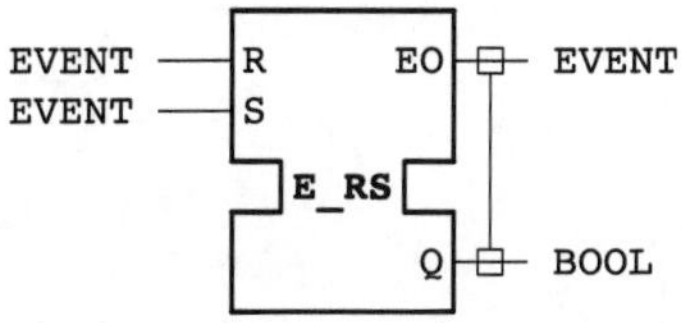

Figure 6.17 Second interface variant of the Event-driven Bistable

The IEC 61499 defines a second event-driven bistable event function block called `E_RS`. Although the interface of this function block features a changed order of its input events (i.e. `R` before `S` as shown in Figure 6.17) it provides the same behaviour as `E_SR`. The reason for having both versions is a consistency with the related IEC 61131-3 types. However, unlike to the IEC 61131-3 types there is no 'dominance' of the events, as only one event is delivered at any time instant (see Section 2.2).

6.2.14 D (data latch) Bistable

The data latch bistable stores the state of a Boolean input `D` on the arrival of a clock input event (`CLK`). If the input state has been inverted, an output event `EO` is produced; the state of the stored Boolean is provided on output `Q` (Figure 6.18).

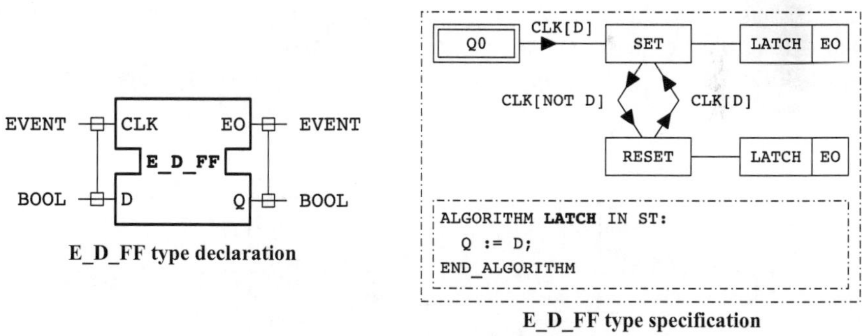

Figure 6.18 D (data latch) Bistable

6.2.15 Boolean Rising Edge Detector

The Boolean Rising Edge Detector produces an output event `EO` when the data input `QI` has changed from `FALSE` to `TRUE` between two occurrences of the input event `E1`. This is best described by the timing diagram in Figure 6.19.

This is a composite function block that uses the functionality of the Data Latch and Event Switch function blocks. Because the Data Latch only produces an output event when the input changes state, an event is only produced when there is a change from `FALSE` to `TRUE`. The Event Switch ensures that only the event for positive transition is selected.

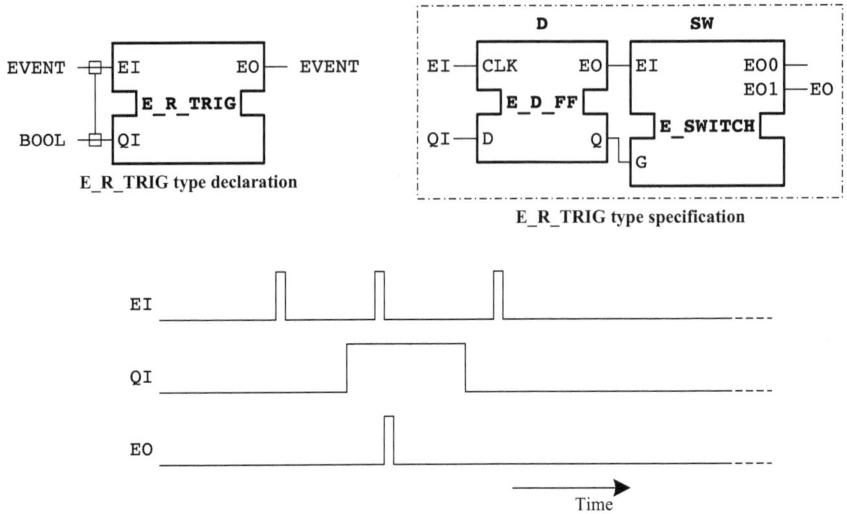

Figure 6.19 Boolean Rising Edge Detector

6.2.16 Boolean Falling Edge Detector

The Boolean Falling Edge Detector produces an output event EO when the data input QI has changed from TRUE to FALSE between two occurrences of the input event E1. This is best described by the timing diagram in Figure 6.20. The Boolean Falling Edge Detector has a similar design to the Boolean Rising Edge Detector.

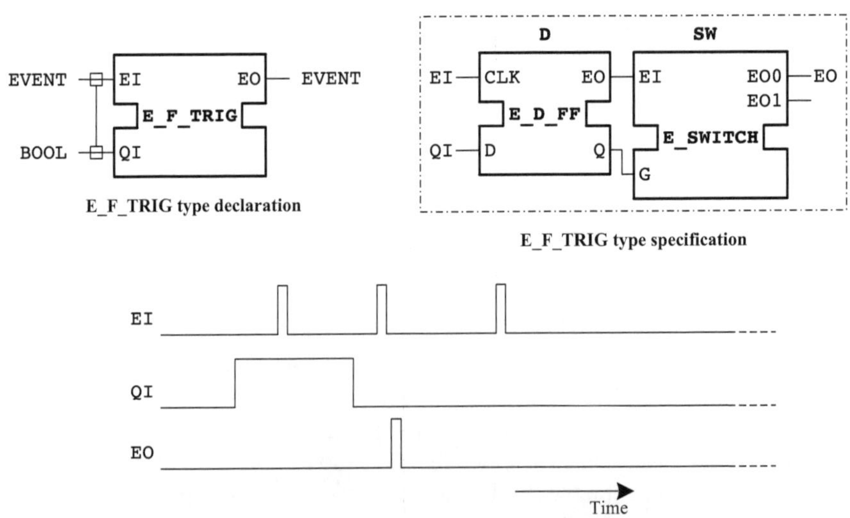

Figure 6.20 Boolean Falling Edge Detector

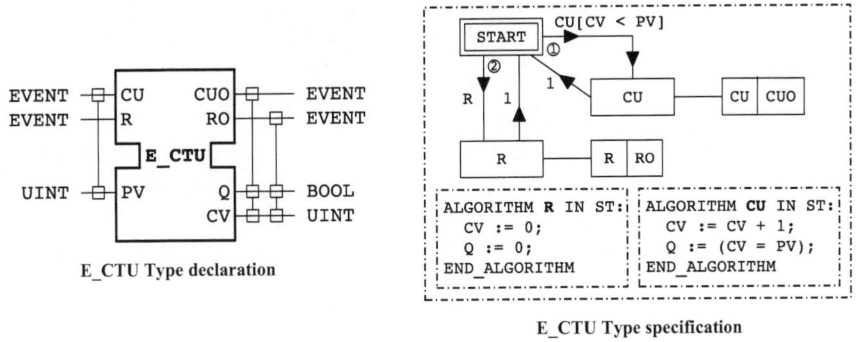

Figure 6.21 Event-driven Up Counter

6.2.17 Event-driven Up Counter

The Event-driven Up counter counts the number of input events arriving on input CU. An output event CUO is produced after each input event has been counted. The count of the number of events is also provided on output CV. When the count has reached the value of PV no further output events are produced and CV remains at the value of PV; the output Q is then set TRUE indicating that the event count is complete (Figure 6.21).

The counter can be reset at any time by a reset event on R. The output Q is then reset to FALSE, the counter CV is cleared to zero and an output event RO is produced.

6.3 Using event function blocks for event flow control

It is envisaged that software tools will provide definitions of these event function blocks in their standard function block libraries. These support the accurate modelling of different kind of tasks for control applications. These range from controlling the basic event flow, to controlling conditional event flow representing small control tasks.

6.3.1 Basic event flow control

We can now review the cyclic event chain example as shown in Figure 6.2 and see how the events can be modelled using the standard event function blocks. Figure 6.22 shows how the event scheme as depicted in Figure 6.2 can be created by inserting standard event function blocks.

The cold start event is provided by an instance of the Restart Event Generator (see FB_E3). The train of events required to maintain the cyclic execution of the two chains of function blocks is provided by an Event Cyclic Generator (see FB_E4). The cold start event and the event train are merged using an Event Merger (see FB_E5). The resulting merged event stream is used to trigger the execution of FB1. The event stream emerging from FB1 is split into two event streams to execute the two function block chains using an Event Splitter (see FB_E1).

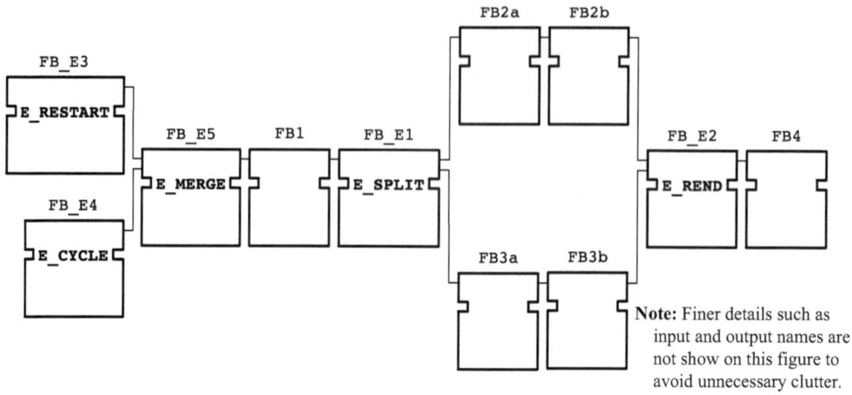

Figure 6.22 Using Event function blocks

The final event stream from the two function block chains, i.e. from FB2b and FB3b are then passed into a rendezvous function block (see FB_E2). This ensures that execution of both function block chains is complete before the execution of the final function block FB4.

It may be argued that the addition of the event function blocks adds extra details that make the function block networks overly complex. However, it should be noted that the resulting diagram is now a formal model that defines the function block execution order precisely and unambiguously independently of the control device that is used.

To simplify function block diagrams, the standard states that some of the commonly used event constructs may be represented by a graphical shorthand notation. Figure 6.23 shows how simple graphical connectors can be used in place of Event Splitter and Event Merger function blocks.

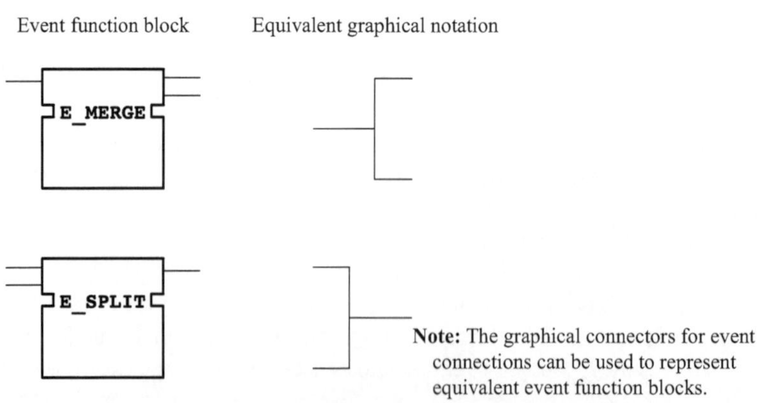

Figure 6.23 Graphical shorthand notation

Note: If graphical connectors are used in this way, software tools should record that each connector represents a particular type of event function block. This is particularly important if the diagram is converted into a textual form.

6.3.2 Conditional event propagation

In the previous example, we have so far considered a simple flow of events through an application (i.e. each function block is triggered once in each cycle). However, often the flow of events and which function blocks are triggered depends not only on events alone but also on the current data values of certain function block outputs. Therefore, several of the IEC 61499 event function blocks (as presented) control event propagation based on their data inputs.

Consider, for example, an application part which processes data received from a communication network using a `SUBSCRIBE` function block. In certain situations there may arise problems during the data reception and the `SUBSCRIBE` function block will inform the subsequent function blocks by setting `QO` to `FALSE`[1]. In these cases, the data processing should not be activated. This can be achieved using an `E_PERMIT` function block as depicted in Figure 6.24. Only if `QO` is `TRUE` and therefore that valid data is available at the `SUBSCRIBE` function block will the `E_PERMIT` function block propagate the event.

While this simple error handling protects the data processing it does not support specific error handling measures. For this it is necessary to activate alternate event paths.

Consider as a second example a control application reading a sensor value using the `IO_READER` function block as described in Chapter 5 and using this sensor value as the basis for its control operation. If the sensor value is correctly read and the IO module has no faults (i.e. `QO` is `TRUE`) the normal control application should be activated. However, if this is not the case (i.e. `QO` is `FALSE`) a failure handling routine, which can for example perform a safe shut-down and notify the user, should be activated. Figure 6.25 shows how this behaviour can be achieved using an `E_SWITCH` function block.

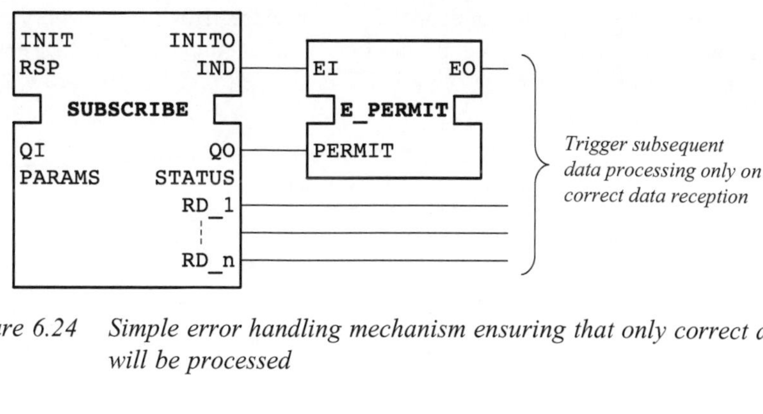

Figure 6.24 Simple error handling mechanism ensuring that only correct data will be processed

[1] According to the general behaviour of service interface function blocks described in Chapter 5.

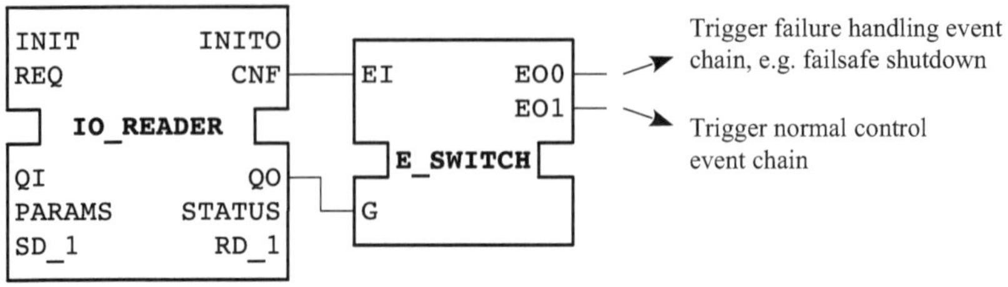

Figure 6.25 Handling errors with two different event chains

6.3.3 Modelling control tasks

Apart from controlling the event flow in applications small control tasks can also be handled with event function blocks. Let us consider a basic control task which is very often used as a starting point for learning PLC programming – the blinking of a light at a given frequency.

At first, we need an element holding the state for the light's value. For this the bistable event function block `E_SR` is best suited. A second main element is the cyclic trigger for turning the light on and off, which is provided by the `E_CYCLE` function block. Finally, we need to toggle the state of the light in each cycle by triggering either `S` or `R` depending on the current state of the light (i.e. `Q`). For this we can again use an `E_SWITCH` function block, which can trigger two different event paths depending on a Boolean value – in our case the state of the light.

The resulting blinking light control application is shown in Figure 6.26. With the `START` and `STOP` events of the `Cycle` function block the blinking can be started and stopped, respectively. By changing `DT` the blinking frequency is adjusted. The outputs of the `State` function block are the update trigger (`EO`) and the value (`Q`) for the light. These can for example be connected to an `IO_WRITER` function block as described in Chapter 5.

As the functionality to oscillate a state (such as for blinking a light) is often required the function block network as presented in Figure 6.26 could be encapsulated in a composite function block for reuse. However, the presented implementation of the blinking lights control is a rather simple one and has several drawbacks. First of all when the blink cycle is stopped (i.e. triggering the `STOP`

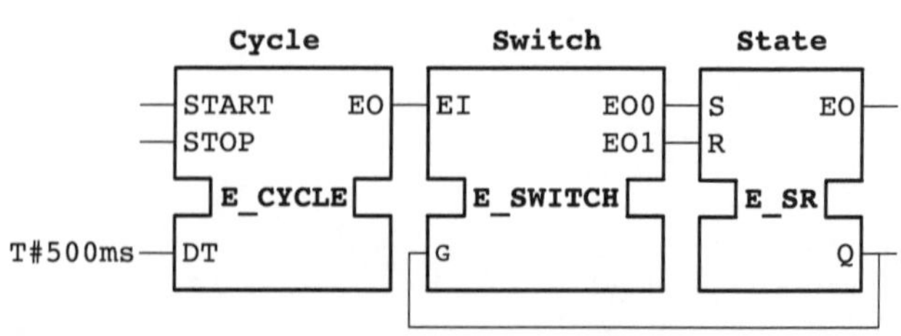

Figure 6.26 Blinking lights control using Event function blocks

event) the state of the light can either be on or off. Secondly, the on and off times are equal. With the use of further event function blocks these limitations can be corrected, which is left as exercise.

Further usage examples of event function blocks in larger applications can be found in the next chapter.

6.4 Summary

We have now covered all of the standard event function blocks defined in IEC 61499 and seen that these can be used to model a wide range of different event schemes. For example, from networks where function blocks are only required to execute once, say, as a response to a system cold start, to complex networks where there are parallel chains of function blocks that require cyclic execution.

To summarise:

- Event function blocks provide a precise definition for event behaviour.
- Events can be created for common situations such as system cold start and warm re-start.
- Event trains can be created where events occur at regular intervals.
- Events can be merged and/or made to 'rendezvous'.
- Events can be produced when Boolean values change state.
- Simple graphical connectors can be used to represent common event function blocks.
- Conditional event propagation can be utilised for error handling paths.
- Event function blocks can also be used for basic control tasks.

Chapter 7
Industrial application examples

In this chapter, we will consider how the IEC 61499 concepts and standard function blocks can be applied to modelling some examples of industrial control systems.

Specifically we will:

- *model a simple temperature control system and its main function blocks,*
- *consider how to design a function block model for a conveyor belt control system,*
- *consider issues related to modelling systems running on Fieldbus devices,*
- *explore some of the special requirements for function blocks used in process control.*

7.1 Overview

Through Chapters 2–6, we have reviewed the concepts and techniques for defining function blocks and applications. This included a discussion on the fundamental elements of the IEC 61499 architecture that covered how applications could be modelled using networks of function blocks. Then we went on to describe concepts such as devices, resources and how applications could be mapped and configured on them. We are now ready to consider how these ideas can be used together to model some typical industrial control problems. This chapter reviews some example applications that have been chosen to illustrate some common control system problems.

There is clearly not space in a book of this format to give detailed model designs. The intention is, therefore, to show how the important aspects of the control system can be modelled and not to spend time on the less important details. Therefore, issues such as error detection and recovery, raising alarms and handling exceptions to normal behaviour are not covered in detail in the following models. However, it is anticipated that the reader should be able to deduce how these models could be extended to cover the range of additional functionality that would be needed in a complete industrial system model.

In the following models, input values depicted on models with a '#' prefix, such as, `#IO_PARAMS` denote configuration constants. It is envisaged that software tools used for creating IEC 61499 models will provide means to assign values to such configuration constants. In some cases, where input values are known to be fixed, literal values are given, e.g. `T#100ms`.

Note: Data types and literal constants are based on the definitions in the PLC software standard IEC 61131-3. A list of data types is given in Appendix A.

7.2 Temperature control example

The first example that we will consider is a simple temperature control system as depicted in Figure 7.1. In this application we have a heat treatment oven, for example, used to cure products made from epoxy resin. The oven is heated by a heater element and its temperature is measured using a sensor. We will assume that the heater is controlled by some form of current regulator, such as a thyristor stack and that the temperature is measured accurately using a thermocouple. For our model, we are just going to consider the high-level aspects of the functionality of this control system, so that finer details such as how to interface with the heater current regulator and sensor are not necessary. We will also assume that issues such as thermocouple linearisation and cold-junction compensation are handled in firmware built into the hardware I/O system. Finally we will assume that the control will all run on a single device within a single resource.

This application has been chosen because it typifies a wide range of industrial instrumentation and control problems. There are many instances where the measurement of a process is taken, in this case by reading the temperature, and then an algorithm is used to adjust some form of flow or process variable. In this temperature control example, we will assume that a simple PID algorithm will be used to monitor the temperature and make appropriate adjustments to the heater current to stabilise the temperature. Similar models could be used for pressure and speed control loops.

To keep the model simple, the oven will be controlled at a fixed temperature; i.e. the set-point temperature is built into the configuration of the control system and is not adjustable. Clearly in a real system, the set-point would normally be adjustable either via an external device (e.g. providing a product specific

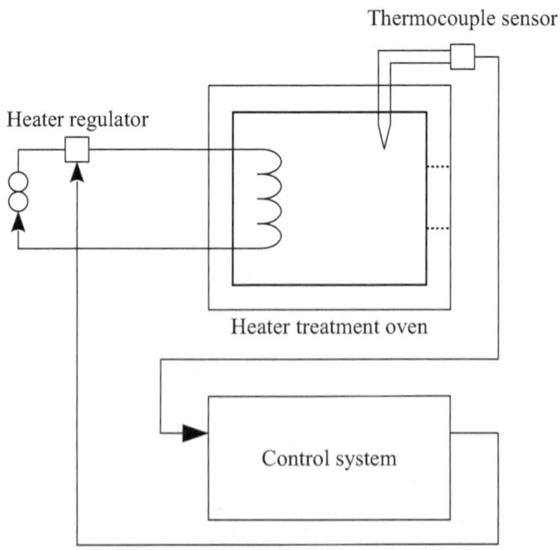

Figure 7.1 Temperature control example

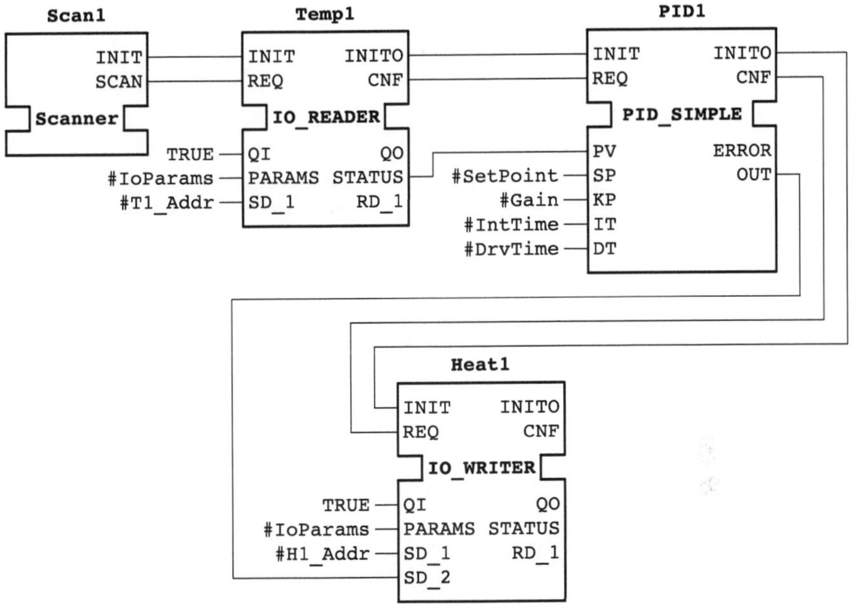

Figure 7.2 Temperature control system function block diagram

temperature profile) or by a control panel. Similarly, the PID tuning parameters such as the proportional range, integral and derivative time constants are assumed to be fixed.

The top-level application model for this control system is shown in Figure 7.2. The model has the following function blocks:

`Scan1` – This is an instance of a `Scanner` function block type and provides the `INIT` and `Scan` events. The `INIT` event triggers the initialisation of the rest of the function blocks that are linked in a chain. The `SCAN` event provides an event every 100 ms to cyclically execute the other function blocks.

`Temp1` – This is an instance of the `IO_READER` function block that is described in Chapter 5. It is used to read in the latest value from the temperature sensor every time it receives an `REQ` input event. The address of the temperature sensor for the device I/O sub-system is specified in the constant `#T1_Addr`. This function block provides a service interface to access I/O values from the device I/O sub-system. When the new value from the temperature sensor has been obtained, the function block triggers its output event `CNF`.

`PID1` – This is an instance of a `PID_SIMPLE` function block and provides the control algorithm that adjusts the heater current. It has two input events `INIT` and `REQ`. The `INIT` event is used to initialise the internal variables used by the PID algorithm. The `REQ` event triggers the execution of the PID algorithm. This causes the PID algorithm to take the latest values for the process value (`PV`) and set-point (`SP`) and calculate the new value for its output to drive the heater. When the PID algorithm has run, output event CNF is produced along with output OUT. Note that the PID algorithm runs every time the Temp1 function block receives a new value for the temperature and has raised event CNF.

Heat1 – This is an instance of the IO_WRITER function block described in Chapter 5 and is used to write the new output value produced by PID1 to the heater current regulator. In fact, like the Temp1, this is a service interface function block that sends values across the resource interface to the device I/O sub-system. In this case, the value placed on input SD_2 is sent to the I/O device as given by the address #H1_Addr. Note that this function block runs every time the PID1 updates its output.

The top-level model shows the main functionality partitioned into function blocks and clearly depicts the execution control. Such partitioning allows design modifications to be readily modelled. For example, an alternative algorithm could be used by simply replacing the instance of the PID_SIMPLE function block with, say, a FUZZY logic control function block.

This is an example of a scanned execution strategy where the function blocks are run at a regular and repeated update rate; this is typical of a closed loop control design. In this case an update rate of 100 ms has been assumed. Using function blocks in feedback control loops is sometimes referred to as 'direct digital control'. From the general theory of digital control there is a theoretical minimum rate at which signals should be sampled to give stable and accurate control. This is typically less than half the time constant of the response of the controlled process to step changes. In practice, establishing the appropriate sampling rate may require detailed information on equipment responses times, process time constants etc. These issues are outside the scope of this book. However, understanding the timing aspects of a controlled process is clearly an important aspect of control system design and modelling.

The Scanner function block type used to create Scan1, is built as a composite function using standard event function blocks as reviewed in Chapter 6, see Figure 7.3. It uses an instance of E_RESTART function block to provide the initialisation event INIT and an instance of the E_CYCLE function block to produce

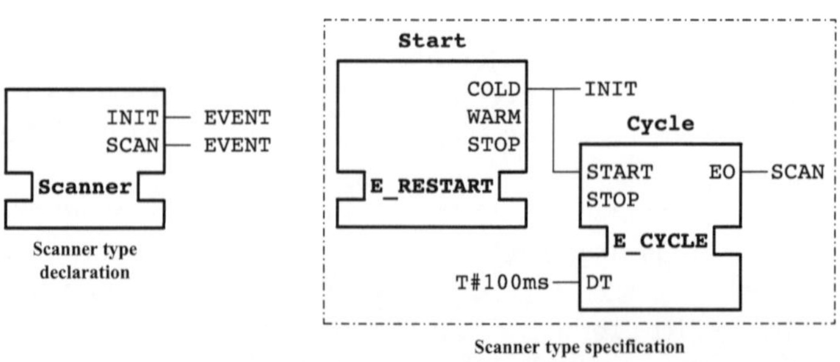

Figure 7.3 Scanner function block

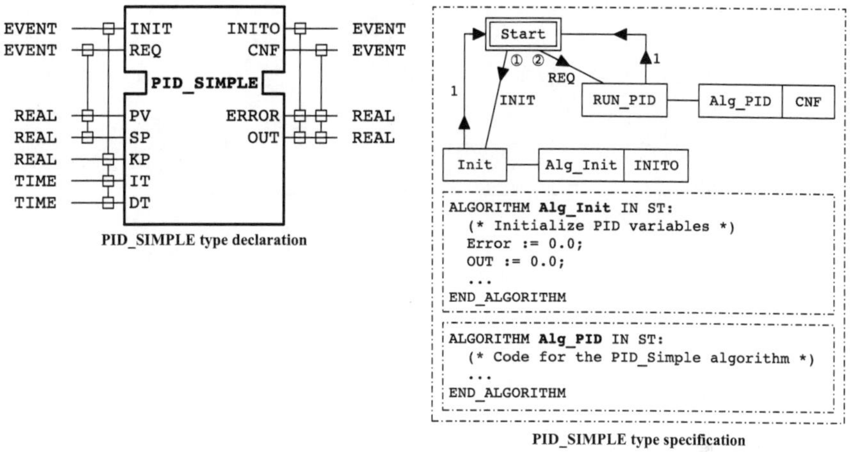

Figure 7.4 PID_SIMPLE function block

the cyclic event `SCAN`. Note that the periodicity of the `SCAN` events is defined by the configuration constant `T#100ms`, i.e. the function block generates an event every 100 ms.

The `PID_Simple` function block type specification is defined as a basic function block and is specified in terms of an Execution Control Chart (ECC) and algorithms written in the IEC 61131-3 high-level language Structured Text (ST). However, note that any language could be used for the algorithm provided the rules for mapping to FB input and outputs are clearly defined – this is discussed in Chapter 3.

Part of this function block's type specification is given in Figure 7.4; this shows the external type specification and the internal type specification. The ECC shows two states `INIT` and `RUN_PID` that exist when this function block receives `INIT` and `REQ` events, respectively. The initialisation algorithm `Alg_Init` runs when the `Init` state is activated, likewise, the PID algorithm `Alg_PID` runs when the `RUN_PID` state is activated. When either algorithm has completed, the ECC always returns to the initial `Start` state.

The function block diagram shown in Figure 7.2 shows the function block network running on a single resource. This same model may be distributed to run on multiple resources by using service interface function blocks, such as `PUBLISH` and `SUBSCRIBE`, to transfer data between resources. For example, this application could be run on two controllers as shown in the subapplication example reviewed in Section 4.1.4.

7.3 Conveyor test station example

In the second application example, we will consider the control of a conveyor used to feed components past a test station as part of a quality acceptance process. The components under test are plastic housings used in the production of an

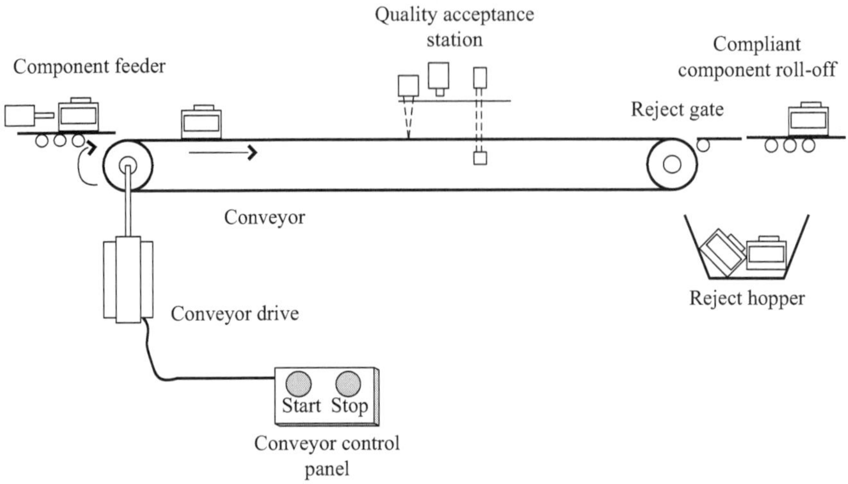

Figure 7.5 Conveyor test station example

electronic product. We will assume that these components have been produced by an injection-moulding machine. Before each component can be used in production it must pass a quality inspection test that checks a number of key parameters including the main dimensions and component identity that is etched into the component surface as a bar-code. A record of each test will be recorded in an external database held on a remote server machine.

The main features of the conveyor are shown in Figure 7.5. This depicts a conveyor belt that is driven by a conveyor drive motor. Components are fed onto the conveyor, one at a time and passed under a quality acceptance station from a component feeder. If the component has compliant quality parameters, i.e. that they are within the accepted tolerance and the component has the correct identity, as read from a component bar-code, the component is passed on to the 'compliant component' roll-off table. On the other hand, if the component fails any of the quality checks, the reject gate is operated and the component is dropped into the reject hopper. As each component leaves the conveyor, a new component is fed on to the belt from the feeder. The conveyor can be started and stopped by an operator from a small control panel that has 'start' and 'stop' buttons.

The top-level function block diagram model for this system is depicted in Figure 7.6. To create this model we have assumed that the control will run on a single resource. Later we will consider how the design could be distributed to run on different controllers. Issues such as dealing with conveyor jamming and other exceptional conditions have not been considered to keep the example simple. We have also assumed that the quality parameters are fixed and can, therefore, be part of the configuration data. In a real system it is likely that these would be provided by an external device or entered from a control panel.

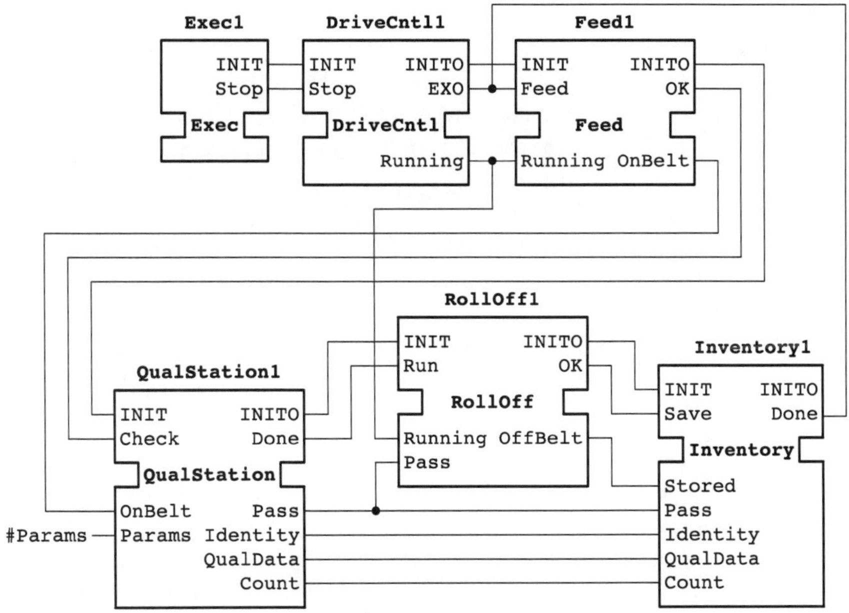

Figure 7.6 Conveyor test station function block diagram

The function blocks that represent the functionality in this design are:

`Exec1` – This is an instance of an `Exec` function block type. This function block fires up a chain of events that keep the whole control system running. It does this by issuing a single initial event `INIT` when the control system is first powered-up. It also produces a `Stop` event when the control system is powered down that causes the conveyor and all equipment to stop.

`DriveCntl1` – This is an instance of a `DriveCntl` function block type. This function block controls the motor drive and its interaction with the control panel; we will look at its internal design later. This function block produces an initialisation event `INITO` when it is first activated, which is used to trigger the initialisation of all the other function blocks. It also produces an output event `EXO` when the motor changes state, i.e. changes from 'stopped' to 'running', or from 'running' to 'stopped'. It also has a `Running` output variable that remains `TRUE` while the belt is moving. The `Running` output variable must be `TRUE` for the feeder and roll-off mechanisms to operate. Note that when `EXO` event occurs and the output `Running` is `TRUE`, this indicates that the motor has started and the belt is moving.

`Feed1` – This is an instance of a `Feed` function block type and is used to control the component feeder. The `EXO` event from the `DriveCntl` is passed to the `Feed1` function block to operate the component feeder. An event at its `Feed` input along with an asserted `Running` input variable causes the `Feed1` function block to push a new component onto the conveyor. If the feed operation is successful, this function block then asserts an `Onbelt` output variable and issues an `OK` event.

`QualStation1` – This is an instance of a `QualStation` function block type and provides an interface into quality test station equipment. Values for the parameters to be tested are passed into this function block on its `Param` input. This function block starts the check of the component dimensions and the scan for the component bar-code when it receives the `Check` input event along with an `Onbelt` variable from the `Feed1` function block. When the component has passed under the test station equipment, this function block issues a `Done` event and sets a number of outputs. These are: a `Pass` output variable that indicates whether the component has passed the quality checks, the component identity, quality data and a count of components tested so far. These values are provided at the output variables `Identity`, `QualData` and `Count`.

`RollOff1` – This is an instance of `RollOff` function block type and provides an interface into the Roll-off table and the reject gate mechanisms. When this function block receives a `Done` event from the test station function block `QualStation1` on its `Run` event input, it reads the values of its `Running` and `Pass` inputs. If the `Running` input is not asserted, this indicates that the belt has stopped and so no action is taken. If the belt is running and the `Pass` input is `TRUE`, the roll-off mechanism is activated to move the component on to the compliant component table. However, if the component has failed the quality tests and the `Pass` input is set to `FALSE`, this function block opens the reject gate and the component is dropped into the reject hopper.

`Inventory1` – This is an instance of the `Inventory` function block type and provides an interface into a database in a remote server to store information from the component tests. When this function block receives an `OK` event from the `RollOff1` function block, indicating that the component has been removed from the belt and a variable indicating that the component is off the belt, it gathers data to be sent to the remote database. The input event `Save` triggers the function block to gather the following information: whether the component has passed the quality test, the component identity, quality data and a count of components tested. This information is sent and stored on a remote server. When the data has been successfully received and stored in the remote database, this function block produces `Done` output event. This is fed back to the `Feed1` function block to trigger the loading of the next component.

Note: The event streams merge where the `Done` event from `Inventory1` is fed back to `Feed1` – shown by a tee-junction. As discussed in Chapter 6, this is graphical shorthand for an instance of a standard event function block `E_Merge`.

This conveyor example is one in which the execution timing of the function blocks is synchronised exactly with the speed of transfer of components along the belt. The `Feed1` function block is only triggered to feed a new component on the belt when the last component has been removed from the belt and its data processed. This contrasts with the scanned execution timing strategy of the previous temperature control example.

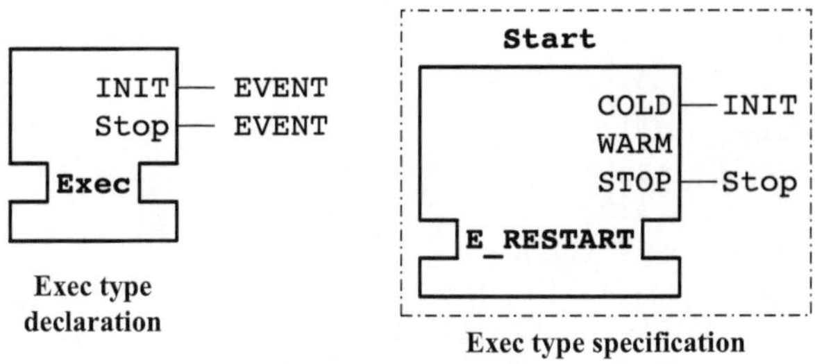

Figure 7.7 Exec function block

The design of `Exec` function block type is shown in Figure 7.7. This is an example where an instance of a standard event function block `E_RESTART` has been re-encapsulated as a new function block type to simplify the external interface.

The `DriveCntl` is an example of a fairly complex component function block as depicted in Figure 7.8. This provides the following functionality:

- It continually reads the values of the start and stop buttons from the control panel, i.e. every 100 ms.
- It sets the motor drive state, i.e. 'running' or 'stopped' and writes this out to the motor drive actuator whenever the panel requests a change of state.
- It produces an initialisation event when the panel and motor I/O have been successfully initialised.
- It produces an `EXO` event when the motor (i.e. the belt) changes state between 'stopped' and 'running'.

Examining the internal design of this function block, we see that the `Cycle` function block, an instance of an `E_Cycle` event function block, produces a stream of events at 100 ms intervals. These events are used to request to read the value of the control panel buttons, i.e. using `Rd_Panel`. For simplicity, it is assumed that the panel buttons can be read as a single Boolean value where `TRUE` means 'run' the motor and `FALSE` means 'stop'.

The 'Run' state is held in an `SR` bistable function block `EnableRun`, an instance of `E_SR` as described in Chapter 6. The 'Run' state is set by the rising edge of output `RD_1` from `Rd_Panel`, which is set to `TRUE` when the start button is depressed. The bistable `EnableRun` is reset when either the stop button is depressed and `RD_1` is `FALSE` or a stop input event is received. Whenever the output `Q` for `EnableRun` changes state, an event is produced on output `EO`. This is used to trigger a request event `REQ` to the `Wrt_Motor` function block to change the state of the motor actuator. When the motor actuator has completed its change of state, i.e. it has switched between 'Off' and 'On', the `Wrt_Motor` function block produces a confirmation event `CNF`. This is used to create the final output event `EXO`.

`Rd_Panel` and `Wrt_Motor` are instances of the service interface function blocks `IO_READER` and `IO_WRITER` as described in Chapter 5 and provide an interface into the device I/O sub-system to read inputs and write to outputs.

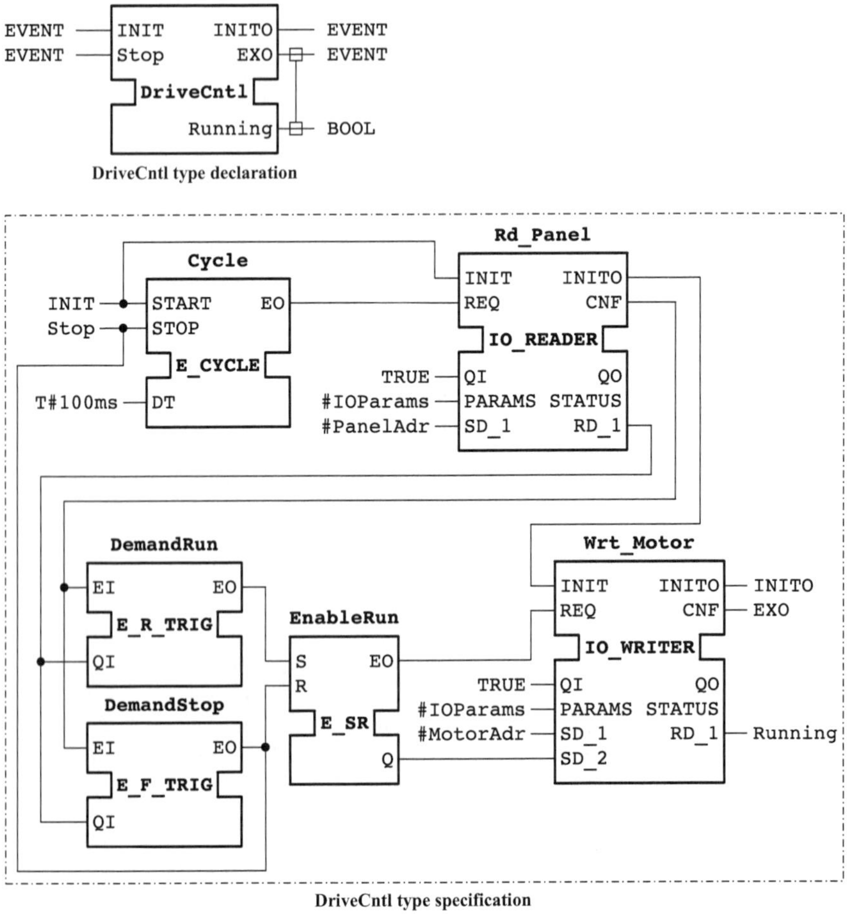

Figure 7.8 DriveCntl function block

We can now outline the designs of the other function block types used in the conveyor test station example:

`Feed` FB type – provides an interface to the feed mechanism; its design will also use instances of the service interface function block types `IO_WRITER` and `IO_READER` to write output values to actuators and read in values from sensors concerned with operating and monitoring the feeder.

`QualStation` FB type – provides interfaces to the quality test station equipment. This design will again use service interface function block types `IO_Writer` and `IO_Reader` to operate and read values from the test equipment and vision inspection units. The interface to the bar-code reader uses serial communications that can be modelled using another form of service interface function block.

`RollOff` FB type – this provides the interface for the roll-off mechanism and reject gate. Again this functionality can be modelled using instances of the service interface function block types `IO_Writer` and `IO_Reader` to control the various actuators and read back the values of sensors.

`Inventory` FB type – this provides the interface to a remote database storing throughput and quality data. Its design will take the input data and transform it into a format suitable for the database. For transferring the data to the database a `CLIENT` function block type can be used.

7.3.1 Distributed system model

We will now consider how the system model for the conveyor test station could be distributed so that it models functionality that runs on different controllers, i.e. devices. Note that if we use the correct IEC 61499 parlance, anything that is capable of supporting a resource is regarded as IEC 61499 device. We will assume that the user panel and motor actuator control is located in one device `Device_A` and the rest of the functionality to handle the component feed mechanism, the test equipment and the roll-off is located in a second device `Device_B`. The two devices as depicted in Figure 7.9 are linked by a local network. The type of network does not impact the design of the top-level model, it is assumed that it provides a reliable communications link between the devices. For this example, we can assume that some form of industrial Ethernet is being used that can reliably send data from one device to another, using a network addressing scheme.

The application design is now split between two IEC 61499 resources located in the two devices. To link the two parts of the application additional instances of `PUBLISH` and `SUBSCRIBE` function blocks have been added as shown in Figure 7.10. These are service interface function blocks that can be used to send data across a communications medium or network. The `PUBLISH` function block `Pub1` sends out data with a particular identifier. This is represented in the design using the configuration constant `#Addr1`. The `SUBSCRIBE` function block, `Sub1` is used to receive data sent out by the `PUBLISH` function block. The `Pub1` and `Sub1` can be considered as an associated pair of function blocks; together these

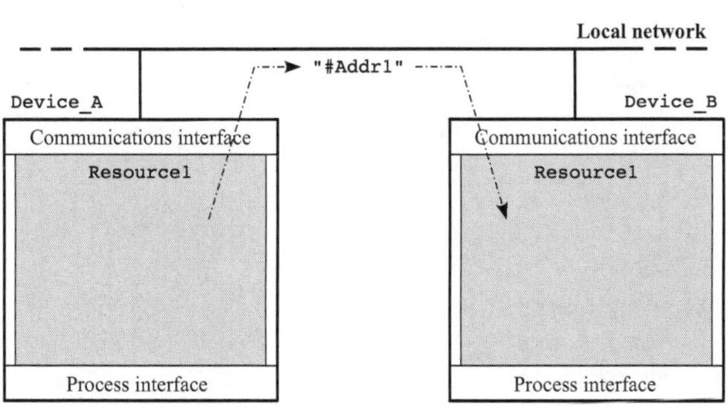

Figure 7.9 Conveyor test station networked devices

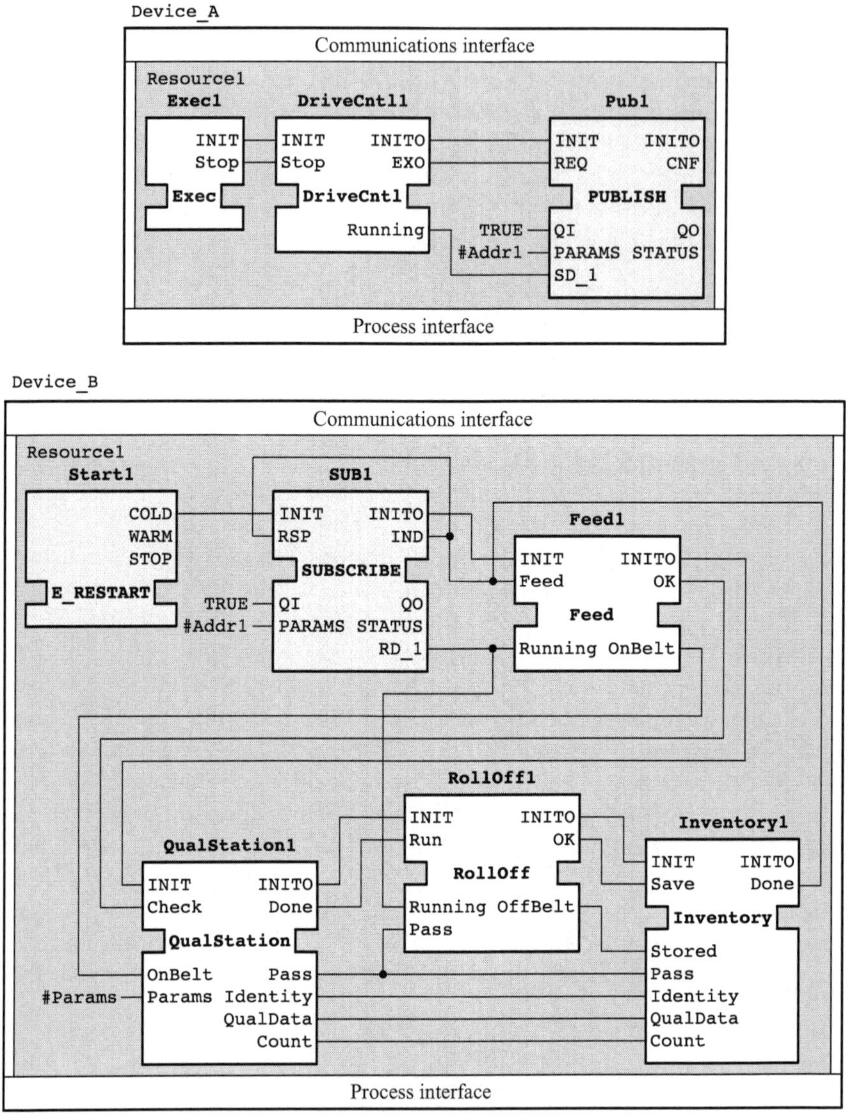

Figure 7.10 Conveyor test station distributed function block diagram

provide the link between the `DriveCntl1` and `Feed1` that existed in the single resource model shown in Figure 7.6.

It is envisaged that tools that create function block networks will be able to automatically insert the appropriate service interface function blocks when applications are broken into parts that are distributed in different resources.

Although the type of network is not important in the high level model, clearly there are performance and reliability issues that will need to be considered.

For example, the additional communications time delay from issuing an REQ event to publish data and receiving the associated RSP event may impact the overall model and system design.

Note: Each resource requires its own initialisation event to start the function block execution. An additional E_RESTART function block has been added to the model in Device_B.Resource1 to ensure that this fragment of the application network is initialised.

Also note that the indication output event IND from the SUBSCRIBE function block Sub1 is fed back to the response input event RSP. This is necessary to inform the SUBCRIBE function block that the data has been processed and therefore ensures that the SUBSCRIBE function block is always 'listening' for the next item of data to be sent by the PUBLISH function block.

7.4 Fieldbus applications

New intelligent plant sensors and actuators are now appearing in the industrial automation arena that combine built-in processing capabilities with standard communications interfaces; such devices are now known collectively as 'Fieldbus' devices – even though they may not all use the same communications standards. In fact, the IEC has been developing a multi-part Fieldbus standard under the designation of IEC 61158 for a number of years. The IEC Fieldbus standard covers the complete ISO communications stack from the physical to application protocol layers. It is intended as global standard that will allow sensors and actuators from different vendors to be compatible at both, the physical level, i.e. by having standard plugs and sockets, and at the software level, i.e. new replacement devices from different vendors can be automatically re-configured to operate in an existing system. It should be possible to build a system using I/O devices from different vendors that interoperate at all levels. For example, it should be possible to remove a pressure sensor produced by vendor A and exchange it for a device from vendor B and the system software should be able to automatically adapt and remain operational.

At the physical layer, the IEC 61158 standard defines a variety of different options for the physical communications media and data rates. There are also wireless and fibre optic options defined in the standard. Unfortunately, this standard has been a long time in development so that a variety of alternative proprietary Fieldbus solutions have been emerging in the marketplace during its development, e.g. using Siemens's Profibus, Allen-Bradley's DeviceNet. Furthermore in the last years several Ethernet-based Fieldbus implementations have been presented, e.g. Siemens's Profinet or Allen-Bradley's EtherNet/IP, which further complicates the situation.

It has been extremely difficult for the IEC to find a suitable compromise solution that is acceptable to all industrial nations and that also fits all types of application domains. For example, field devices used in a hazardous area of a petrochemical plant will need to be intrinsically safe, i.e. designed for low voltage and

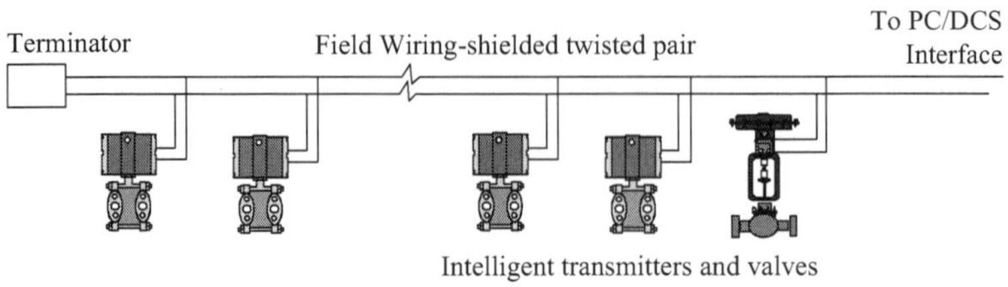

Figure 7.11 Fieldbus concepts based on the IEC 61158-2 standard

low energy to remove risks of gas ignition; these devices will also need to react at relatively high speed. On the other hand, field devices used by water utilities, for example, to monitor water treatment plants, will generally need to be low cost and operate over longer distances at lower data rates. Having so many conflicting requirements has resulted in a wide range of different options and solutions being defined for Fieldbus communications standard – so that interoperability of different devices at the physical level may not always be possible. Currently there is also a variety of proprietary implementations of the Fieldbus concepts that allow devices either from a single vendor or group of vendors to interoperate. Therefore the IEC 61158 and partnering standards (e.g. IEC 61784) define different implementation types based on the proprietary Fieldbus implementations.

Although a single standard for all industrial field devices may never be achieved, the issues regarding how to design and model systems built from networks of intelligent devices is still relevant. IEC 61499 concepts can apply to both IEC Fieldbus as well as the proprietary Fieldbus devices.

When using Fieldbus, devices are connected to a common communications media so that data can be transferred between any two or more devices. The layout of physical wiring arrangement of Fieldbus devices is not determined by the application. Devices can generally be cabled together in way that is most convenient to the installer, such as in a daisy chain that minimises point-to-point cabling, see Figure 7.11. The linkage between the processing elements within each device is achieved by software by routing data between devices in such a manner as to solve a particular application.

Figure 7.12 shows part of an application created by linking instrumentation in different Fieldbus segments. This is part of a causticising process used to make wood pulp for the paper manufacturing. Each instrument provides a 'resource' that is capable of executing function blocks. Each Fieldbus device may either have a built-in repertoire of function blocks or may be installed with new function blocks; for example new function block type definitions can be downloaded and stored in some form of non-volatile storage within an instrument. Applications are created by configuring the software links between Fieldbus devices and, in some cases, between controllers. Where controllers are connected these will be typically Distributed Control System (DCS) work stations but other types of controllers such as PLCs may be used. With small applications, clusters of

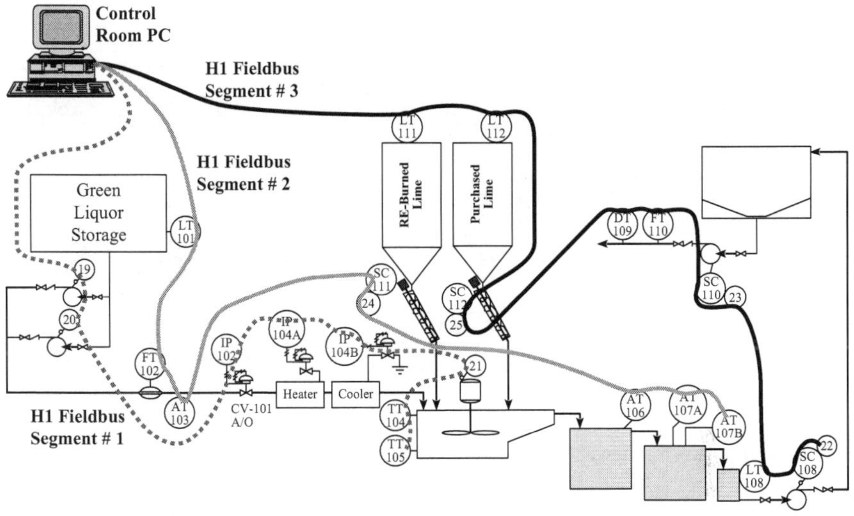

Figure 7.12 Distributed instrumentation using Fieldbus

Fieldbus devices may be configured into simple applications without the need for a main controller. For example, a simple control loop could be configured by linking an analogue input instrument, such as a temperature controller, that has an on board PID function block, with an intelligent valve that has analogue output function block.

The top-level functional control requirements can be expressed in terms of the function blocks that exist within the distributed instruments as shown in Figure 7.13. Function blocks within clusters of instruments concerned with a specific aspect of the control can be linked into a network. The example shows part of a cascade and ratio control loop used in the causticising process. The main advantage of using Fieldbus devices is that the logical connections between function blocks within the instrumentation is de-coupled from the physical location of the instruments. It is also necessary to identify the timing relationships between the various function blocks. A large number of function blocks can be scheduled to run concurrently if they have no inter-dependencies. Part of the execution strategy for this example is shown at the bottom of Figure 7.13.

The next stage is to use IEC 61499 models for the various function blocks to create an IEC 61499 application. This will include the addition of service interface function blocks to provide the communications links between the IEC 61499 Resources that exist in the different instruments.

There is a proposal to develop a range of standard Fieldbus function blocks that can provide the main types of functionality required in general process control applications; these are listed in Figure 7.14. Of course, additional specialist function blocks will also be required in other control domains and instrument vendors will also be free to develop their own function blocks to encapsulate proprietary algorithms or to interface with new types of instrumentation.

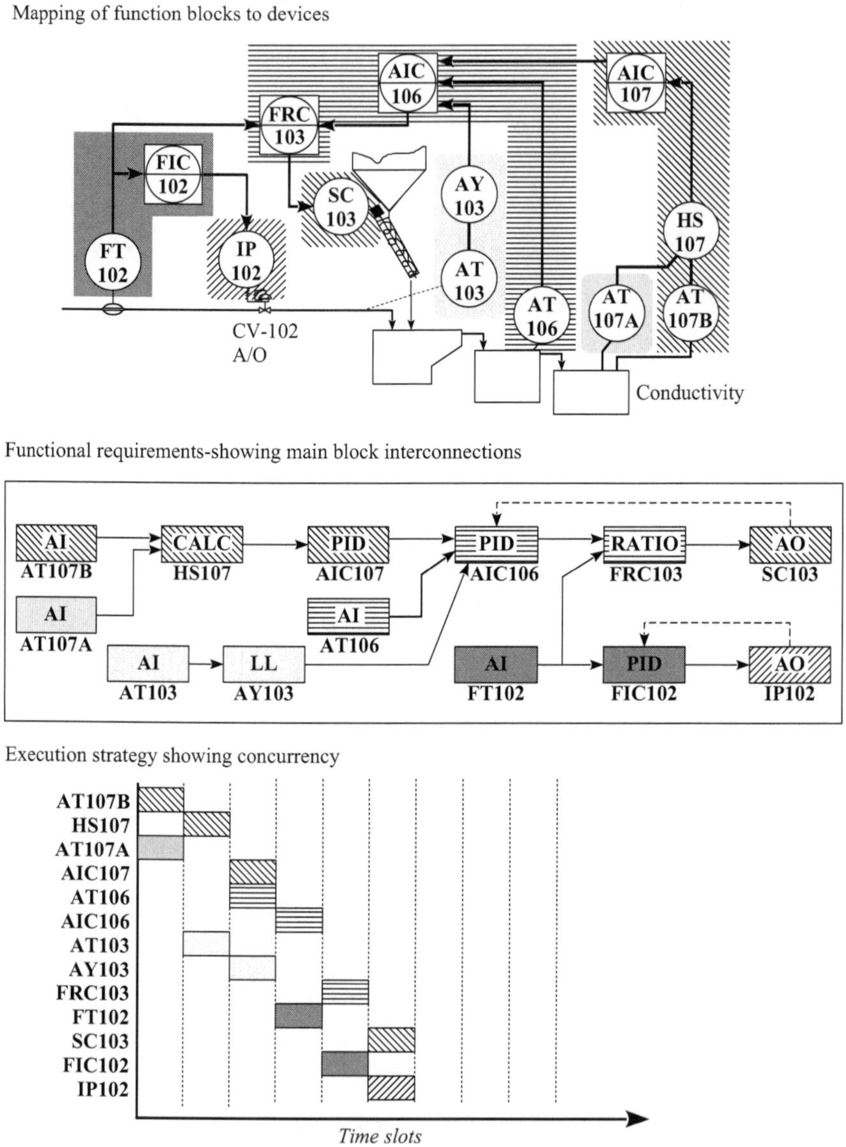

Figure 7.13 Mapping Fieldbus function blocks

7.4.1 Analogue input function block example

We will now consider a typical Fieldbus function block for handling analogue inputs and see how the IEC 61499 concepts can be applied to model its behaviour. The main features of the control of an analogue input are shown in Figure 7.15. This function block provides an interface for a range of different transducers, such

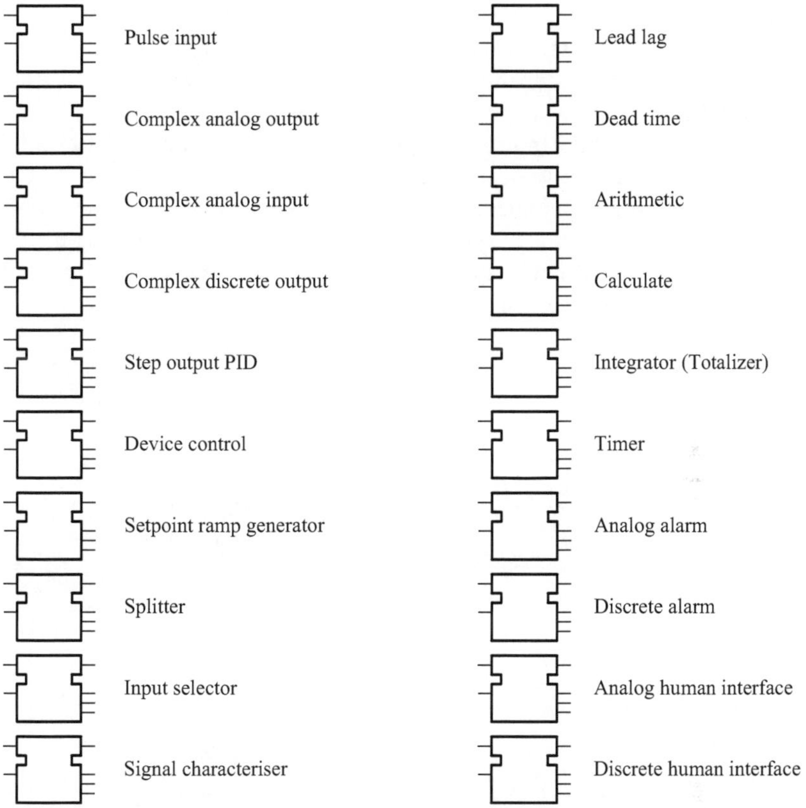

Figure 7.14 Fieldbus function blocks

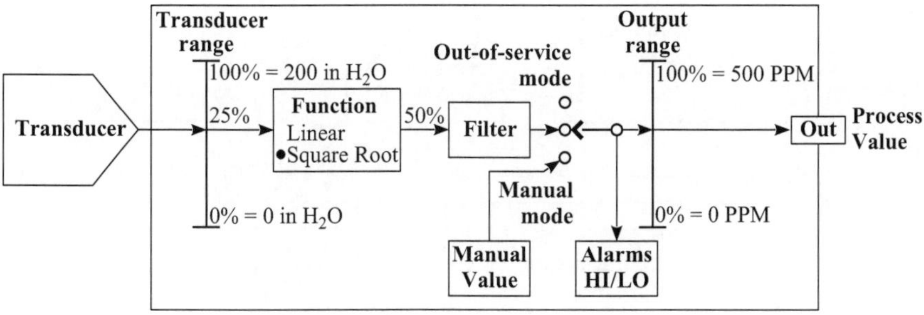

Figure 7.15 Analogue input functionality

as for flow, pressure, temperature, water quality (impurity measured as Parts Per Million (PPM)) and so on.

The raw input measurement from the transducer is first converted into a working range or span. This generally requires calibration to set the maximum and minimum values. The calibrated value from the transducer is then passed through some form of linearisation function to convert the values to engineering units. The form of linearisation function will depend on the type of transducer. In this example, we assume that the transducer is measuring some form of water impurity that requires a square root function to convert the measured value into a measurement in PPM.

The value is then filtered to remove short measurement spikes that may occur due to noise from the transducer. Finally the value is converted into an output range. There is also a requirement to set an alarm output when a certain threshold value is reached. There is also provision for modes for out-of-service operation and manual operation where a manually supplied value can be used instead of the value read in from the transducer. This is provided for commissioning and testing purposes.

Figure 7.16 depicts the `AnalogInput` function block type that models the main parts of the behaviour of this analogue input. Additional functionality that would be required for an analogue input in a real system, such as, provision for further high and

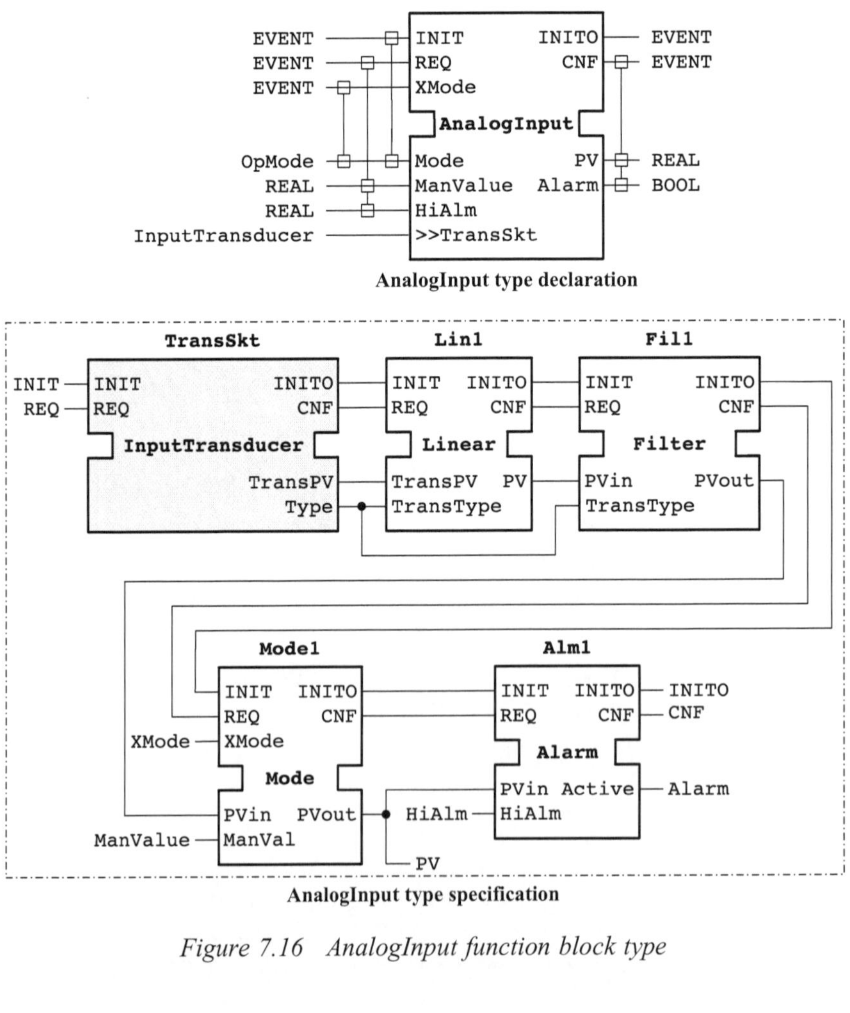

Figure 7.16 AnalogInput function block type

low alarm levels and alarm hysteresis, and setting calibration scales for the Process Value (PV) span range and so on, are not shown to simplify the example.

This function block uses the adapter concept, as discussed in Chapter 4, to allow it to work with different types of transducer; i.e. the functionality to handle the interfaces with different types of transducer is contained in external function blocks that are connected to the analogue input.

The external interface of the `AnalogInput` function block is shown at the top of Figure 7.16; there are three input events – `INIT` to initialise the function block, `REQ` to cause a new process value to be read from the transducer and `XMode` to change the operating mode.

The function block also has the following inputs:

- `Mode` an enumerated variable that defines the operating mode, i.e. `O_S` for out of service, `MANUAL` and `AUTO`.
- `ManValue` input gives the process value to be used when the analogue input is switched to manual mode.
- `TransSkt` a transducer socket that requires an adapter connection from a transducer function block providing a plug of adapter type `InputTransducer`.
- `HiAlm` input defines the alarm level for the high alarm threshold.

On the output side, the function block has two output events – `INITO` that occurs when the function block with its connected transducer has been initialised and `CNF` that occurs when the function block has read a new process value `PV` from the transducer. If the process value exceeds the high alarm value, the Boolean output `Alarm` is set `TRUE` when the `CNF` output event occurs.

The internal specification of this function block is depicted at the lower part of Figure 7.16. The shaded function block `TransSkt` is the transducer socket and depicts the interface that must be presented by an external transducer function block via an adapter interface.

The transducer produces two outputs `TransPV` and `Type` that are passed into the linearisation function block of type `Linear`. The linearisation function block provides an algorithm that is appropriate for the type of transducer and takes the raw transducer process value `TransPV` and converts it into engineering units. This value is then passed through a `Filter` function block. The `Filter` function block also has an input to read the transducer type. This is required in order to modify the filter characteristics to match the transducer type. The `Filter` function block produces the final output process value `PV` that is then passed into the `Mode` function block for handling the changes in operating mode.

The `Mode` function block provides mode switching between: (a) 'out-of-service mode' where the process value is not updated, (b) 'auto mode' where the process value is derived from the value read in from a transducer and (c) 'manual mode' where the process value is derived from the manual input `ManValue`. The internal algorithm tests the value of the enumerated input `Mode` when the change mode event `XMode` arrives and selects the appropriate state. The execution control chart for the `Mode` function block is shown in Figure 7.17. This example assumes that the mode can be changed at any time. In a complete industrial design for an

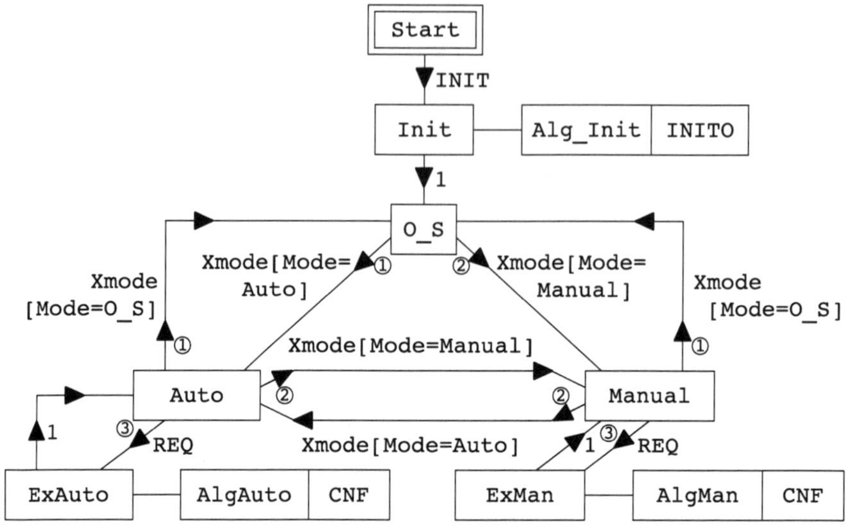

Figure 7.17 Mode execution control chart

analogue input, there may be additional constraints concerning when the mode can be switched; for example, for safety reasons there may be an inhibit signal to prevent switching to and from the auto mode. This could be handled by adding further inputs to the Mode function block and changing the expressions for the transitions between the mode states.

Finally, an Alarm function block provides the test to compare the process value with the high alarm value and produces the Alarm output when the process value exceeds the high alarm threshold given at its HiAlm input.

Note: The input events INIT and REQ are passed directly into the transducer socket. This means that these events will be passed into the transducer function block that is connected externally to the AnalogInput function block via the adapter interface.

Figure 7.18 shows a small application that has been created using two instances of the AnalogInput function block, Press1 and Flow1. These have been characterised by connecting function blocks for pressure and flow transducers, i.e. Trans1 and Trans2, to create analogue input values for pressure and flow, respectively. The transducer function blocks are connected to the AnalogInput function blocks using an adapter connection. This allows each AnalogInput function block to have access to the transducer function block via the transducer function block's external interface. In effect the transducer function block provides the TransPlg of the adapter type InputTransducer shown in the AnalogInput type specification in Figure 7.16. The transducer function block modifies the AnalogInput's behaviour to be 'specific' for a particular transducer type.

In this example application, the two analogue measurements for pressure and flow are passed to a function block Panel1 that handles a small display panel. The Scan1 function block provides initialisation event INIT to initialise the AnalogInput function blocks and their transducers and also an execution event

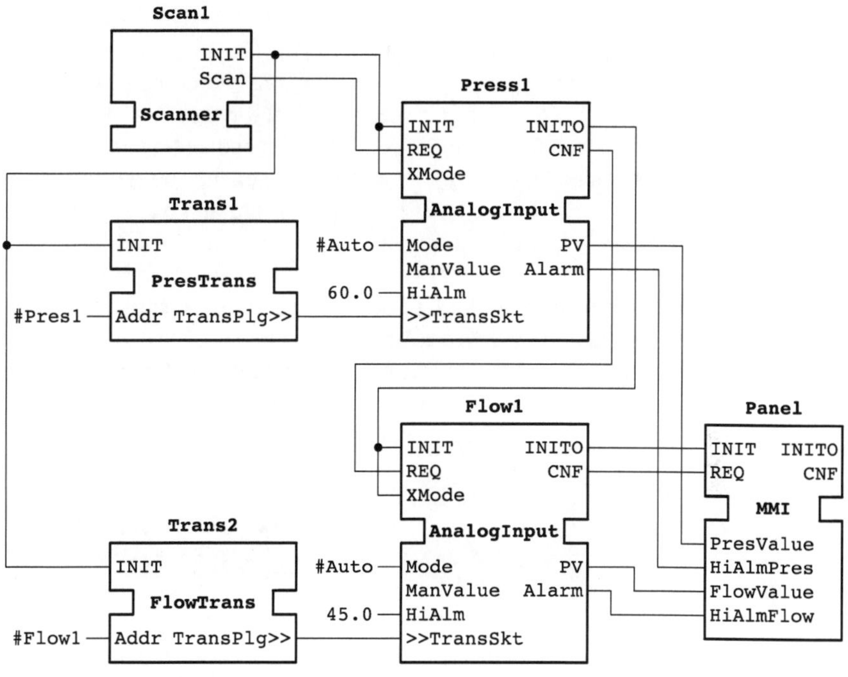

Figure 7.18 Using the AnalogInput function block

Scan to ensure that the analogue input process values are repeatedly updated at a regular rate, such as, every 200 ms. To simplify this example, the mode facilities are not used. In fact the mode of both function blocks is set to Auto when the AnalogInput function blocks are initialised.

Note that each transducer function block receives two initialisation events – one received on its external interface on event input INIT. This is used to cause the function block to read the transducer I/O address and initialise its interface with the hardware. A further initialisation event is received across the adapter interface on event input INIT of the TransSkt socket as shown in Figure 7.16; this is used to initialise the interface between the AnalogInput function block and the transducer.

7.4.2 Further generalisation of the analogue input function block

The analogue input example showed how the handling of the input value can be decoupled from the specific input transducer while at the same time providing a rich interface using the adapter concept. However, the implementation of the Analog-Input function block as shown in Figure 7.16 still contains a dependency on the specific input transducer type. This dependency is given by the TransSkt's Type output needed for adjusting the behaviour of the Filter and Linear function block to the specific input transducer. This dependency has the drawback that every time a new filter, linearisation or input transducer type is needed adjustments within the Linear and Filter function blocks are necessary. As the number of types increases the maintenance and extensions will become more difficult.

A point which has been neglected in the implementation of the `AnalogInput` function block so far is the specification of parameters for the `Linear` and `Filter` function blocks. In the given implementation two options would be possible. The first option is to add all configuration parameters for each type at its interface and only provide the parameters for the type used. The second option is to provide generic parameter interfaces (e.g. array of `ANY`) and depending on the type being used provide the necessary parameters. Both options are not very intuitive or convenient to use and are also difficult to maintain and extend.

In order to improve this situation the `AnalogInput` function block can be further generalised by transforming the `Linear` and `Filter` function blocks into adapters. The change to the `InputTransducer` interaction is that we will use plugs for both in order to indicate that we require an implementation for this functionality. The `GenAnalogInput` function block type definition showing this is presented in Figure 7.19. The two new plugs can be seen at the bottom right side of the function

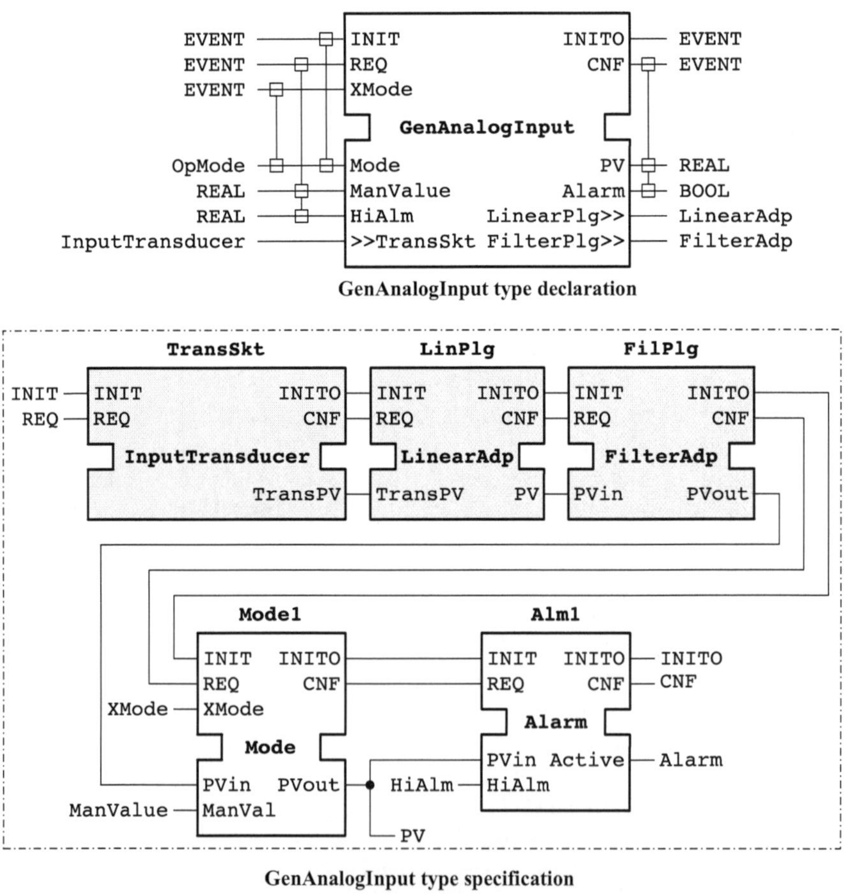

Figure 7.19 Generic AnalogInput function block type

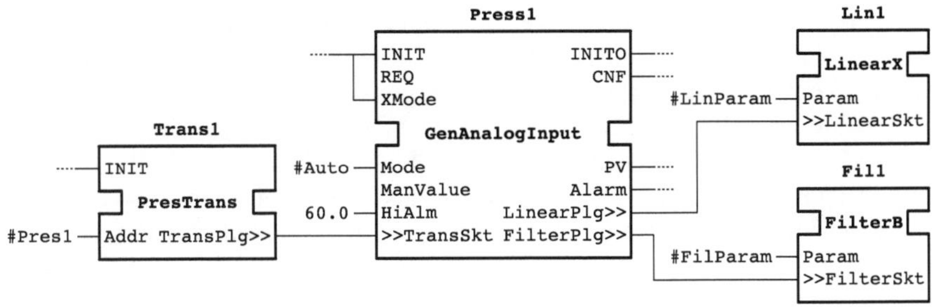

Figure 7.20 Using the generic AnalogInput function block

block's interface. The function block network within the `GenAnalogInput` function block has the same structure as the non-generic case presented in Figure 7.16. The main difference is that the `TransSkt`'s `Type` output has been removed.

When using the `GenAnalogInput` function block in addition to the `InputTransducer`, function blocks providing sockets for the `Linear` and `Filter` plugs are also required. Figure 7.20 shows how the pressure measurement part of the example application needs to be adapted when using the `GenAnalogInput` function block. The two additional function blocks `Lin1` and `Fil1` provide the sockets for the two plugs of the `GenAnalogInput` function block. Each of the two function blocks has its own set of input values allowing their parameters to be specified easily. By using different function block types the filter or linearisation can be easily adapted.

The only drawback however is that the `GenAnalogInput` function block is not self-contained any more. It can only be used when all its sockets and plugs are connected. Depending on the application it has to be considered whether this additional effort is justified. The generalisation approach with adapters can bring great benefits as showed in this example. Especially when it is required that the coordination logic can be separated from the algorithms and the algorithms need to be adjusted to application specific requirements.

Summarising this generalisation approach, the `GenAnalogInput` function block defines the general behaviour regarding execution flow while the specific implementations providing the plugs and sockets (i.e. the input transducer, linearisation and filter algorithms) define and detail the specific actions in this design.

7.5 Concluding remarks

As a final note in this chapter, the approach taken when modelling or designing systems using the IEC 61499 function block concepts will be the same irrespective of the types of systems and the types of devices that are used. As discussed in Chapter 1, the first stage in designing any function block based system is to consider the top level Functional Requirement Diagram. This outlines the functionality in terms of the main function blocks and interconnections. The next stage is to apply the IEC 61499 concepts to decompose the function blocks of functionality into networks of function blocks with data and event flow connections. This can include hierarchical decomposition where larger function blocks are specified as subapplications or composite function blocks made up from smaller function blocks. The function blocks can then be distributed to run on processing resources that are interconnected by various networks and communication links. Here, the term processing resources applies to any device that is capable of running function blocks, e.g. PCs, Programmable Logic Controls, DCS out-stations and intelligent devices and instruments. In some cases, the required functionality may already be built into a particular type of instrument. For example, the model may require a cascade PID controller – so the functionality expressed in the IEC 61499 model may reflect the behaviour of existing devices. Having the complete IEC 61499 function block system model with information on all function blocks' internal specifications, algorithms and timings, it is then possible go on and verify the design and simulate the system behaviour. The IEC 61499 function block model also provides an implementation independent view of the system. This could be particularly useful when it is necessary to formally verify and validate a system design; for example, as would be necessary for a safety-related system.

7.6 Summary

This chapter has considered a range of industrial applications that demonstrate the features of the IEC 61499 function block standard. This has shown that function block model is very flexible and can be easily adapted to model different types of system arrangements. This has ranged from systems that require scanned execution strategies to those where the function block execution is directly related to the way the plant is behaving and from systems that run on a single resource to those that are distributed across resources connected in a network.

It has also been shown that the function block model can be applied to Fieldbus devices where software functionality is distributed in instrumentation linked via communications networks.

The adapter concept is also very useful for creating generic function blocks where behaviour is characterised by the connection of external function blocks using a 'plug' and 'socket' interface. This concept is particularly useful for function blocks dealing with families of input and output devices such as analogue inputs and outputs.

Chapter 8

Epilogue

In this final chapter, we will review how IEC 61499 standard is likely to develop in the future and its possible impact on the design of systems and support tools.

This chapter will review:

- *current limitations of Part 1 of the IEC 61499 standard,*
- *IEC 61499 Part 2 that covers 'Engineering Task Support',*
- *compliance with IEC 61499 as defined in IEC 61499 Part 4.*

8.1 Current status of IEC 61499

We have now covered the main concepts defined in the first part of the IEC 61499 Function Block standard. This book has focused more on providing an overview and showing how the standard can be applied rather than exploring some of the more complex details of its formal definitions. One area, which has not been explored in any great detail, has been the textual syntax that is defined within Part 1.

The textual syntax is defined formally in terms of 'production rules' that can be used by compiler writers to build software tools for processing function block designs. This is an important feature of the standard because it means that designs that are produced graphically can be stored in a textual form that is portable to different support tools. It should be emphasised that all function block type specifications, applications and subapplications have two equivalent forms. They can be defined either graphically, which has been the main form used in this book, or they can be defined in a concise textual form. Tools are already available that can automatically convert between graphical and textual forms.

To recap, IEC 61499 Part 1 defines an architecture, models and definitions that allow function blocks and applications to be defined in a precise form with behaviour that is well understood. However, to create and manage systems based on function block concepts requires further definitions and concepts; this is the domain of Part 2 of the IEC 61499 standard.

Part 2 covers 'Engineering Task Support' and includes the following topics:

- Specification of function blocks in a form that can be manipulated by software tools.
- The functional specification of resource and device types.

- The specification, analysis and validation of distributed Industrial Process Measurement and Control Systems (IPMCS).
- The configuration, implementation, operation and maintenance of distributed IPMCS.
- The exchange of information among software tools.

At the time of producing this edition of this book, the parts 1, 2 and 4 have just gone through their first maintenance phase and have been republished as second edition. During this maintenance phase mainly ambiguities regarding the execution model of function blocks have been resolved. These ambiguities were heavily discussed in the scientific community and were definitely a hindering point for widespread industrial adoption. The new execution model not only resolves this issue it is also much easier and resource efficient to implement.

The IEC 61499 working group also published a Part 3 of IEC 61499, which provides 'Tutorial Information' on applying the IEC 61499 concepts. Because of different reasons, this part was withdrawn in 2008, but its revision and republication is currently under reconsideration. This part includes examples of how software tools can be used to create, edit and manipulate function block designs, applications and configure function block oriented systems.

As discussed in Chapter 1, IEC 61499 function blocks can be used to provide a view of how the functionality of a system is distributed and how algorithms are executed. It does not yet cover other areas of the system design lifecycle. For example, IEC 61499 does not cover how requirements are captured and used to create an initial functional design. It also does not cover issues, such as, the definition of low level communication protocols that are needed to load, initiate and configure function blocks – this has to be part of domain specific standards that apply the IEC 61499 function block concepts. IEC 61499 is a generic standard and therefore does not attempt to define domain specific function blocks, so the standard does not contain definitions for common function blocks such as analogue inputs, PID algorithms. Such definitions should be developed in domain specific standards. It is an IEC objective that the IEC 61499 will be referenced in future standards that deal with function blocks which may range from domain standards for process control, building management to safety systems.

A further apparent limitation of IEC 61499 is the fact that it does not give any details on how function blocks are actually implemented. How does the resource actually schedule event delivery and function block algorithms? The standard defines the requirements and rules but does not give details on the implementation. This issue is outside the scope of IEC 61499 and must be dealt with in domain specific standards. For example, we would expect Fieldbus standards to cover the implementation of function block scheduling in a Fieldbus device. On the other hand, devices used in building management systems may be based on different hardware and use different software operating systems and therefore will require different techniques for scheduling function blocks. Several different implementation approaches fulfilling the needs and requirements of different application domains and use-cases have been implemented and evaluated so far. A good overview of these approaches can be found in [20].

Issues related to how function blocks are compiled, downloaded and stored within a device are also related to the implementation and so are not specified in the IEC 61499 high level functional model. Compliance profiles as described later in this chapter are concerned with addressing these issues.

So although IEC 61499 currently does not cover the complete system design lifecycle and has not yet been applied in domain specific standards, such as, for function block designs for process control, we may ask, is it useful? The answer must be yes – because it provides for the first time a precise methodology for modelling function block behaviour. This will allow models to be ported between software tools and, in the future, for tools to be developed to validate and simulate IEC 61499 system models. Throughout this book we have discussed IEC 61499 as a system modelling technique rather than as a design methodology. This is because, per se, it cannot be used to express all the information required for a complete system design. However, as a model it can be used to express the most important aspects of a control system as it shows the relationships between executable algorithms, the events that trigger their execution and their timing. Using IEC 61499 it is now possible for software tools to simulate the execution of system models and prove that the models are correct.

Several software tools and execution environments – commercial as well as free – are now available. These allow the full description of the control systems with the models defined in IEC 61499 and use these models on execution environments for performing the real-time control of the system. In several applications from different domains, including building automation, laboratory automation, process control or machine control the proposed advantages of IEC 61499 have been demonstrated.

However, some shortcomings in the modelling process and elements have also been identified. Some of these shortcomings were already considered in the maintenance phase for the second edition. For large changes or extensions to IEC 61499, the IEC maintenance process is not very well suited. Therefore, the IEC 61499 working group decided to collect these changes in a separate new part – Part 5 'Proposed Extensions'. This allowed on the one hand, a timely maintenance for the preparation of the second edition and on the other hand gave the possibility of a Publicly Available Specification (PAS) phase for collecting feedback from users. The proposed extensions currently consists mainly of extended modelling elements allowing a more efficient modelling process and also a more complete modelling of the overall distributed IPMCS.

8.2 Engineering Support Task

Part 2 of IEC 61499 is concerned with the 'Engineering Support Task' and defines the main capabilities of software tools for handling function block–oriented systems. A key requirement addressed in Part 2, which applies to any IEC 61499 compliant engineering support system, is the capability to create and store library elements and also to be able to exchange library element designs between different support systems.

Part 2 defines a 'library element' as a collection of declarations that apply to data types, function block types, adapter types, resource types, device types and system configurations. In effect, this includes all IEC 61499 elements that can be specified as defined in Part 1. Any software tool for handling function blocks must clearly be able to create, edit, store and retrieve all of the different types of IEC 61499 library elements. The standard states that software tools should be able to produce library elements such as function block type specifications in a file format that is implementation independent. In other words, means should be provided in all software tools that are compliant to IEC 61499 to be able to produce files containing library element definitions, such as function block specifications, that can be exchanged between tools and systems from different vendors.

The main features of an IEC 61499 engineering support tool are summarised as follows:

- To store, retrieve and exchange library elements.
- To implement particular library elements, e.g. to create an instance of a function block in a particular resource.
- To browse and display the declarations of selected library elements.
- To create and modify the definitions of library elements, e.g. to open a function block type definition and modify its internal specification.
- To validate the correctness of library element definitions, e.g. to be able to check a composite function block specification and ensure all internal connections are between points of compatible data types.
- To produce compilable forms of function blocks suitable for either committing to firmware or downloading into devices. Also to be able to produce connections between function block instances in a form that can be established across a network.

An important requirement for providing these functionalities is the availability of a standardised vendor neutral file exchange format. The ability to exchange software designs between different vendor's products has been a serious limitation of the IEC 61131-3 PLC software standard because IEC 61131 failed to include a definition of a file exchange format. It has not been possible for graphical PLC software designs using IEC 61131-3 languages to be exchanged between different PLC software support tools.

With the definition of the XML-based exchange format by the PLCopen in 2009 [21] the situation finally has improved. As described in Section 2.6, IEC 61499 provides in Part 2 the definitions for describing all library elements in an XML-based exchange format. This format captures not only the structural information like the textual syntax but also stores graphical layout information. Therefore, this file exchange format provides the infrastructure required for an IEC 61499 engineering support tool.

8.3 Compliance to IEC 61499

As already stated the intention behind IEC 61499 has been a generic architecture and models for distributed IPMCS. Therefore, it only describes the implementation requirements but not implementation details. These issues are outside the scope of IEC 61499 and must be dealt with in domain specific standards. However, already during the PAS phase of IEC 61499 Part 1 and 2, the need for describing the implementation details of IEC 61499 implementations has been identified. In order to homogenise these descriptions, the so-called 'Compliance Profiles' have been introduced.

8.3.1 IEC 61499 compliance profiles

The IEC 61499 compliance profiles specify how the features of IEC 61499 Part 1 and 2 are implemented. This is in clear contrast to the compliance statements of IEC 61131-3, which listed only the features implemented by a vendor. IEC 61499-4 defines the scope, structure and required content of a compliance profile. Figure 8.1 depicts the core scope of compliance profiles. This is to support portability, inter-operability and configurability of devices and software tools from different vendors.

The outline and content of a compliance profile is loosely based on the structure of IEC standards and should be as follows:

TITLE – compliance profile titles should have the form 'IEC 61499 Compliance Profile for "xxx"'. The 'xxx' is free form text describing the scope of the compliance profile.

Foreword – describes the origin as well as the organisations or people involved in the development of the compliance profile.

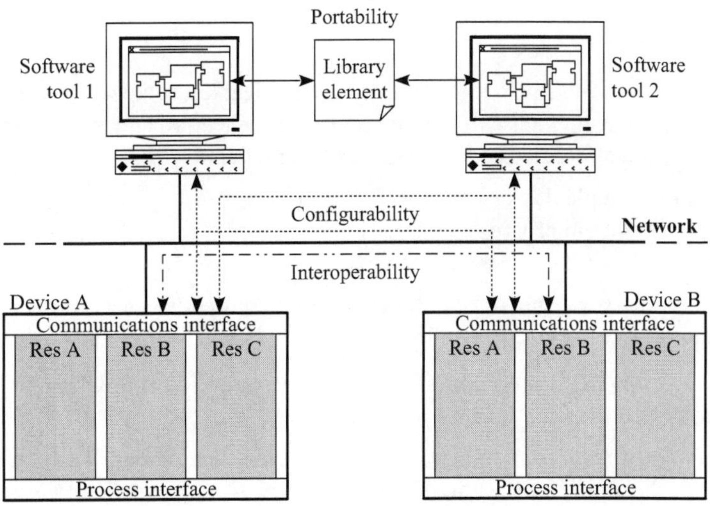

Figure 8.1 Scope of an IEC 61499 compliance profile

Scope – defines the intended usage, the functional scope considering portability, interoperability and configurability and any additional requirements or restrictions on the definitions of IEC 61499 Part 1 and 2.

Normative documents – references any further normative documents required for the compliance profile according to the rules in Part 2 of the ISO/IEC Directives.

Terms and definitions – stated according to the rules given in Part 2 of the ISO/IEC Directives.

Portability provisions – defines the extensions to software tools to enable the creation of files for library element definitions according to IEC 61499-2, as well as extensions to software tools for reading files containing library element definitions compliant to IEC 61499-2.

Interoperability provisions – defines means to support communication between devices and resources, using the definitions and semantics of the communication function blocks as described in Section 5.4. Depending on the communication system used, not all communication models may be supported by a compliance profile.

Configurability provisions – defines means by which software tools can configure devices (e.g. instantiate function blocks, create connections, configure parameters). Different configurability classes may be defined in order to support different levels of configurability (see below).

Test requirements – specifies any tests or test procedures for ensuring compliance with the compliance profile.

Annexes – Compliance profiles can have normative or informative annexes. Normative annexes need to be fulfilled by compliant implementations. Annexes need to clearly indicate if they are normative or informative.

Note: Appendix B gives an overview of an example compliance profile.

8.3.2 Device classes

The IEC has recognised that it would be impracticable to expect all system products to be fully compliant with all facets of the architecture and models defined in Part 1. It has therefore defined the following compliance classes:

Class 0 – Simple devices
Class 1 – Simple programmable devices
Class 2 – User re-programmable devices

A brief summary of the device functionality required to meet these different compliance classes is given as follows:

Class 0 Simple devices: For a device to be compliant with Class 0 it should provide basic functionality to support:

(a) the creation of connections between function blocks,
(b) the ability to start and stop function blocks and applications, and
(c) to be able to support an external query (e.g. arriving across a network) to provide definitions of function block external interfaces.

The query service provides a means to explore the types of function blocks with their input and output data types that exist within the device. Such information is required by a network configuration tool, which is setting up connections between function blocks in different devices.

Class 1 Simple programmable devices: This device class has additional functionality beyond Class 0 that includes:

(a) the ability to create new function block instances as well as connections,
(b) the ability to delete instances and connections,
(c) the ability to start and stop function blocks and applications, and
(d) the ability to support queries about data types, function block types and connections.

Devices in this class will typically have a predefined set of function blocks, for example, held in firmware; i.e. there is no support for changing on-board function block type definitions.

Class 2 User-programmable devices: The final compliance class provides all the functionality defined in Part 1. In addition to the functionality of Class 1, these devices have the ability to create new data types and new function block types – i.e. to be able to create completely new configurations with different function block definitions. This device class also supports a richer range of queries to provide access to all the features of function blocks and applications. Such devices must allow new function block type definitions to be downloaded across a network from a configuration tool.

8.4 Large-scale industrial applications

The use of IEC 61499 concepts in the design of large-scale industrial applications is now becoming a reality. With IEC 61499 providing a comprehensive architectural framework for the design of distributed systems and with extensive provision for reusable components (at different levels of granularity), designs based on this standard can be scalable, readily reconfigurable and, if required, fault tolerant.

A major challenge to the design of any large-scale application is to ensure the design is 'vendor independent'; i.e., it is not locked into using equipment and software solely from a single supplier. Concepts from IEC 61499 (an open standard) encourages the development of system designs based on the interoperability of disparate devices.

Real-world applications for which IEC 61499 is currently being considered include [24]:

Smart grid: The German DKE[1] Standardisation Roadmap lists the IEC 61499 standard as a promising approach for the implementation of control algorithms used for smart grids. A smart grid is a new methodology for

[1] DKE – Deutsche Kommission Elektrotechnik Elektronik Informationstechnik (German Commission Electrical Engineering Electronics Information Technology).

controlling and managing large electrical grids using information technology and advanced communications. It enables an electrical grid to respond rapidly and appropriately to changes in both energy usage and generation, by consumers and suppliers respectively. A key objective is the improvement in the efficiency of the energy distribution through the application of reactive control strategies. Such strategies could be readily encapsulated in IEC 61499 function blocks. It is envisaged that smart grid concept could cover large geographical areas requiring the close integration of millions of sensors (e.g. smart meters in consumer premises) and actuators (e.g. switchgear in sub-stations).

Airport baggage handling systems: An airport baggage handling system is a typical example of a large-scale industrial automation project that requires the close integration of hundreds of different sensors, actuators and controllers from disparate vendors. The design requires modular approach that supports use of 'smart' devices with embedded intelligence that can be interconnected via various forms of communication bus (e.g. Fieldbus). IEC 61499 supports a fault tolerant approach to the system design, so that operation can be maintained, albeit at a reduced performance, when different parts of the system become inoperable, e.g. when a single baggage conveyor fails, baggage can be re-routed via alternative conveyors.

Smart buildings: Large buildings and offices require complex control systems to manage functionality, such as, for HVAC (i.e. heating, ventilation and air conditioning), lighting and security. The instrumentation of a building is truly distributed and requires the integration of a range of different sensors (e.g. for building temperature, light levels, room occupancy) and intelligent devices, such as, controllers for heating, lighting and lifts. The Austrian company nxtControl is an example of a company that is involved in marketing software for a range of building automation projects using concepts from IEC 61499, which will help with this complex integration task.

8.5 Summary

As this is the end of the last chapter, we have reached a point where we have considered all the main aspects of the new IEC 61499 function block standard. We have seen how the various elements defined in Part 1 fit together to create a system architecture for function block execution. The standard covers the key features of the function block specification including formal definition of its external interface including the inputs and outputs of the function block and their association with input and output events.

We have seen that the standard includes a special form of state machine, the execution control chart, to define how events and algorithms are related within a function block. There are also special forms of function block for dealing with the interfaces to external services, such as accessing device I/O values or communicating with other function blocks in other resources.

In Chapter 7, we reviewed how these concepts can be applied to some industrial applications. We have seen that the function block approach is particularly applicable to Fieldbus devices. In this final chapter, we have reviewed IEC 61499-2 and considered some of capabilities of engineering tools that are required to manage and create function block–oriented system models. This included a discussion on the use of a file exchange format based on XML, which allows to capture the function block designs structurally as well as their graphical layout information in a standardised way.

We also considered how vendors can specify their implementation details ensuring interoperability, portability and configurability of different devices and software tools using the rules for compliance profiles defined in Part 4 of IEC 61499. Finally, we gave an overview of several large-scale industrial application domains, where IEC 61499 is currently implemented successfully, showing its advantages.

Appendix A

Common elements

This appendix outlines the common elements used in IEC 61499 that are based on definitions from the IEC 61131-3 PLC Software Standard. This appendix is provided as a brief overview – the formal definitions for all of these elements are given in Part 3 of the IEC 61131 standard [11].

A.1 Character set

All textual information should use a restricted set of letters, digits and characters. The standard requires that only characters from standard ISO 646 'Basic code table' are used.

Optionally lower case letters can be used but they are not significant in language elements, e.g. `Ramp1` and `RAMP1` are treated as identical. However, lower case letters are still significant if used to define printable strings such as `'Load Job 1a'`. Upper and lower case letters can be used within comments.

Note: In IEC 61499 and IEC 61131-3, identifiers are case insensitive but language keywords are case sensitive and should always be in upper case.

A.2 Identifiers

Identifiers are used for naming different elements within the IEC languages, e.g. for naming variables, new data types, function blocks, applications. An identifier can be any string of letters, digits and underlines provided:

- The first character is not a digit.
- There are not two or more underline characters together.

The standard provides options for supporting identifiers with lower case letters, and identifiers with a leading underline character. There can be no embedded spaces. The following are acceptable identifiers:

```
FBMAIN_P1  INPUT_3PV   aEvent1  _APP1    ✓
```

The following are illegal:

```
FB__PV    EX 3PV   1EXpa
```

The standard states that the first six characters should be tested for uniqueness. The following two identifiers may therefore be regarded as identical on some systems:

```
ABX456_XY   ABX456_69
```

Consequently, to avoid the possibility of ambiguous identifiers, when developing software that may be ported to different systems, it is important that at least the first six characters are always unique.

A.3 Comments

Various length comments from short to multi-line can be inserted in any textual definitions of IEC 61499 library elements. Comments can be placed wherever it is acceptable to insert one or more spaces.

Comments are framed by the characters `(* ....... *)`.

Examples are:

```
(* Event input qualifier *)
(***************************)
(* Service initialisation   *)
(* Comment
   on several lines
*)
```

A.4 Data types

The IEC data type range includes floating point numbers for arithmetic computations, integers for counts and identities, booleans for logic operations, times and dates for timing and managing batch systems, strings for holding textual information and bits, bytes and words for low level device operations.

Literals for each data type are included in the following descriptions. Note that a literal is a constant that defines how a given value for a particular data type is represented, e.g. integer literals are `12, 343, −3, 0`. Where the data type of a literal may be ambiguous, the data type can be prefixed, e.g. `INT#12, LINT#342`.

A.4.1 Integer

Table A.1 Integer data types

IEC data type	Description	Bits	Range
SINT	Short integer	8	−128 to +127
INT	Integer	16	−32768 to 32767
DINT	Double integer	32	-2^{31} to $+2^{31}-1$
LINT	Long integer	64	-2^{63} to $+2^{63}-1$
USINT	Unsigned short integer	8	0 to 255
UINT	Unsigned integer	16	0 to $2^{16}-1$
UDINT	Unsigned double integer	32	0 to $2^{32}-1$
ULINT	Unsigned long integer	64	0 to $2^{64}-1$

Integer literals

These can be represented using decimal values, e.g.

`-123 0 +463 23_123`

As binary values, i.e. base 2 values, e.g.	`2#1111_1111` (255 decimal) `2#0000_1000` (8 decimal)
As octal, i.e. base 8 values, e.g.	`8#377` (255 decimal) `8#020` (16 decimal)
As hexadecimal, i.e. base 16 values, e.g.	`16#FF` (255 decimal) `16#A0` (160 decimal)

Note: The standard allows the underline character '_' to be inserted into numeric literals to aid readability, otherwise it has no significance.

A.4.2 Floating point (REAL)

Table A.2 Floating point (REAL) data types

Data type	Description	Bits	Range
`REAL`	Real numbers	32	$\pm 10^{\pm 38}$ (Note 1)
`LREAL`	Long real numbers	64	$\pm 10^{\pm 308}$ (Note 2)

Note 1: REAL values have a precision of 1 part in 2^{23}.
Note 2: LREAL values have a precision of 1 part in 2^{52}.

The format is defined by standard IEC 559; this is identical to the commonly used IEEE floating point number format.

REAL literals

These can be represented using normal decimals and can be distinguished from integer literals by the presence of a decimal point. For large and very small values, the exponential notation can be used giving the power of ten by which the number can be multiplied; either uppercase 'E' or lowercase 'e' can be used to denote the ten's exponential value, e.g.

`20.123, +12_133.21, -0.0001598, -2.65E-10, 0.9E10, 0.6276e+14`

A.4.3 Duration (TIME)

Table A.3 Time data types

Data type	Description	Bits	Usage
`TIME`	Time duration	Note	Storing the duration of time after an event

Note: The length and precision of this data type are implementation dependent. The `TIME` data type is used to store durations, i.e. in days, hours, minutes, seconds and milliseconds.

TIME literals

There are two forms of time literals provided, short form and long form. The short form is concise and fairly readable but where improved readability is required, the long form can be used.

In both forms the following letters are used:

d = days, h = hours, m = minutes, s = seconds, ms = milliseconds

Short form examples:

`T#16d3h3s`	defines 16 days, 3 hours and 3 seconds
`T#3s30ms`	defines 3 seconds and 30 ms

Long form examples:

`TIME#7d_10m`	defines 7 days, and 10 minutes
`TIME#16d_5h_4m_4s`	defines 16 days, 5 hours, 4 minutes and 4 seconds

The last field of a time literal can also be given in a decimal format.

Examples:

`T#12d3.5h`	defines 12 days, 3.5 hours
`T#10.12s`	defines 10.12 seconds

A.4.4 Dates and times of day

Table A.4 Dates and times of day data types

Data type	Description	Bits	Usage
DATE	Calendar Date	Note	Storing calendar dates
TIME_OF_DAY or TOD	Time of day	Note	Storing times of the day, i.e. real time clock
DATE_AND_TIME or DT	Date and time of day	Note	Storing the date and the time of day

Note: The length of this data type is implementation dependent.

Dates and time of day literals

There are two forms of time of day and date literals, short form and long form. The short form is concise and fairly readable but where improved readability is required, the long form can be used.

Table A.5 Dates and times of day literal prefixes

Data type	Short form	Long form
DATE	D#	DATE#
TIME_OF_DAY	TOD#	TIME_OF_DAY#
DATE_AND_TIME	DT#	DATE_AND_TIME#

Examples of DATE literals are:

`D#2001-06-10`	10th June 2001
`d#2004-01-13`	13th January 2004
`DATE#2002-10-15`	15th October 2002

Examples of TIME_OF_DAY literals are:

`TOD#10:10:30`	10 o'clock, 10 minutes and 30 seconds
`TOD#23:59:59`	1 second to mid-night
`TIME_OF_DAY#05:00:00.56`	0.56 seconds past 5

Examples of DATE_AND_TIME literals are:

`DT#2002-06-12-15:36:55.40`	12th June 2002, at 15 hours, 36 minutes and 55.4 seconds
`DATE_AND_TIME#2008-02-01-12:00:00`	Mid-day on 1st February 2008

A.4.5 Strings

Table A.6 String data type

Data type	Description	Bits	Usage
`STRING`	Character strings	8 per character	Storing textual information

STRING literals

A string literal defines a number of characters enclosed in single quotes:

Examples of string literals are:

```
'Batch AX201_65'
'ProductID'
```

A.4.6 Double-byte character strings

Table A.7 Double-byte character string data type

Data type	Description	Bits	Usage
`WSTRING`	Character strings	16 per character	Storing unicode textual information

WSTRING literals

A double byte string literal defines a number of ISO/IEC 10646 characters enclosed in double quotes:

Examples of double-byte character string literals are:

```
"Batch  AX201_65"
"User Name:"
```

Note: These examples are in English and could as well be represented in single byte strings. Typically double byte strings are used when strings in different languages need to be handled.

A.4.7 Bit strings

Table A.8 Bit string data types

Data type	Description	Bits	Usage
`BOOL`	Bit string of 1 bit	1	Digital, logical states
`BYTE`	Bit string of 8 bits	8	Binary information
`WORD`	Bit string of 16 bits	16	" "
`DWORD`	Bit string of 32 bits	32	" "
`LWORD`	Bit string of 64 bits	64	" "

Bit string literals

Bit string literals used to define the contents of bit strings having multiple bits. The same representation as defined for integers, e.g. `146, 2#1101` (value 1101 in binary), `16#FFAC` (value FFAC in hexadecimal), can be used.

A.4.8 Boolean

The boolean data type is used to define the state of boolean variables, such as those associated with digital inputs and outputs.

Table A.9 Boolean data type

IEC data type	Description
`BOOL`	Has two states `FALSE`, equivalent to 0, and `TRUE` equivalent to 1.

A boolean variable can be assigned logical values `FALSE` or `TRUE`. Alternatively the logical values 0 and 1 can be used.

Boolean literals

`FALSE` and `TRUE` can be used to define the literal values of booleans and are treated as reserved language keywords.

```
FALSE    0
TRUE     1
```

A.5 Use of generic data types

The elementary data types that have similar properties are organised in a hierarchy as shown in Figure A.1. Generic data types that start with the prefix `'ANY_'` can be used for describing variables in function blocks where there is support for

```
ANY
 ├─ ANY_ELEMENTARY
 │   ├─ ANY_MAGNITUDE
 │   │   ├─ ANY_NUM
 │   │   │   ├─ ANY_REAL
 │   │   │   │   └─ REAL LREAL
 │   │   │   └─ ANY_INT
 │   │   │       └─ SINT INT DINT LINT USINT UINT ULINT UDINT
 │   │   └─ TIME
 │   ├─ ANY_BIT
 │   │   └─ BOOL BYTE WORD DWORD LWORD
 │   ├─ ANY_STRING
 │   │   └─ STRING WSTRING
 │   └─ ANY_DATE
 │       └─ DATE_AND_TIME DATE TIME_OF_DAY
 └─ ANY_DERIVED
```

Figure A.1 Hierarchy of IEC data types

overloaded inputs and outputs. Overloaded refers to the ability of a variable to be used for different types of data. For example, a function block input `RATE` that can accept all types of numeric values could be defined using the data type `ANY_NUM`. This would imply that the input can be used with any data type from the hierarchy that comes under the generic data type `ANY_NUM`.

A.6 Initial values of elementary data types

IEC 61131-3 standard states that all variables have a known initial value. By default all variables are initialised to zero or in the case of strings to an empty null string. Dates default to the value of `D#0001-01-01`. Alternative initial values can be defined for derived data types.

A.7 Derived data types

New data types can be defined from the elementary data types for use within function block specifications. A new data type is declared by framing the definition with `TYPE` and `END_TYPE`. For example, all values for pressures could be held in a derived data type called PRESSURE, which is based on the REAL data type but with an initial value of 1.0.

The type can be declared as:

```
TYPE
  PRESSURE :REAL := 1.0;
END_TYPE;
```

A.7.1 Structured data types

New composite data types can be defined by creating a structure using existing data types. A structure is declared by framing the definition with STRUCT and END_STRUCT. For example, the data from a pressure sensor may consist of the following information:

- the current pressure as an analogue value,
- the status of the device, i.e. operating or faulty,
- the calibration date,
- the maximum safe operating value,
- the number of alarms in the current operating period.

This can all be defined in a single structure as a new data type PRESSURE_SENSOR:

```
TYPE PRESSURE_SENSOR:
  STRUCT
    INPUT:        PRESSURE;
    STATUS:       BOOL;
    CALIBRATION: DATE;
    HIGH_LIMIT:   REAL;
    ALARM_COUNT: INT;
  END_STRUCT;
END_TYPE
```

A.7.2 Enumerated data types

Enumerated data types allow different states of a value to be given different names. For example, all function blocks within a particular system may have a number of different operational modes. A data type suitable for variables holding function block modes would be:

```
TYPE BLOCK_MODE:
   (OUT_OF_SERVICE,MANUAL,AUTO,CALIBRATE);
END_TYPE
```

Note that variables of type BLOCK_MODE can only be assigned values from the list of named states, i.e. as given in the enumerated list.

A.7.3 Sub-ranges data types

A sub-range can be specified that limits the range of a value for a particular data type. For example, variables used to store voltages for controlling a set of DC motors may be restricted to values +12 and −6 V. A suitable data type for `MOTOR_VOLTS` would be:

```
TYPE
   MOTOR_VOLTS: INT(-6..+12);
END_TYPE
```

A.7.4 Array data types

An array can consist of multiple elementary data types or other derived data types. For example, consider a variable required to hold a set of operating pressures for various points around a steam vessel. Suitable data types would be:

```
TYPE VESSEL_PRESS_DATA :
  ARRAY[1..20] OF PRESSURE;
END_TYPE

TYPE VESSEL_MATRIX :
 ARRAY[1..3,1..4] OF VESSEL_PRESS_DATA;
END_TYPE
```

Note: `PRESSURE` is also a derived type. `VESSEL_MATRIX` is a two-dimensional array of 3 by 4 elements where each element is itself a `VESSEL_-PRESS_DATA` array.

Notes: The use of array and structured data types can greatly simplify the connections between function blocks allowing multiple values to be passed between blocks using a single connection. When composite data is passed into a function block input, all data within the composite value must be valid at the time when the input's associated `WITH` input event occurs.

Appendix B

IEC 61499 Compliance Profile for Feasibility Demonstrations

This appendix outlines as an example an IEC 61499 compliance profile, designated: the 'IEC 61499 Compliance Profile for Feasibility Demonstrations'. This appendix is provided as a brief overview – the full compliance profile can be found at http://holobloc.com/doc/ita.

The 'IEC 61499 Compliance Profile for Feasibility Demonstrations' was the first compliance profile developed and is still the most used. Currently, most IEC 61499 implementations comply at least in part to this compliance profile. Originally, the profile was developed within the 'Holonic Manufacturing Systems' project to show the inter-vendor feasibility of IEC 61499. The original version was defined by the companies Rockwell Automation, Softing AG and Profactor GmbH. Currently it is hosted and maintained by the company Holobloc Inc.

Note: Compliance profiles are typically developed and maintained by organisations not under direct control of the IEC.

The scope of this compliance profile is to provide reference definitions for all three aspects (i.e. portability, interoperability and configurability) of IEC 61499 compliance profiles. How these three aspects are defined in the 'IEC 61499 Compliance Profile for Feasibility Demonstrations' is outlined in the following sections.

B.1 Portability provisions

In order to comply to the portability conditions, software tools need to be able, for all the library elements supported, to generate and to correctly parse and interpret the syntax according to the XML DTD definitions in Annex A of IEC 61499-2. Furthermore if the software tool uses files as an interchange form the following file suffixes are defined:

- '.dtp' for data type definitions
- '.fbt' for function block type definitions
- '.adp' for adapter type definitions
- '.res' for resource type definitions
- '.dev' for device type definitions
- '.sys' for systems

B.2 Interoperability provisions

The compliance profile defines interoperability for devices utilising Ethernet communication with the protocols TCP[1] and UDP[2]. Hereby TCP is used for the bidirectional communication model (i.e. `CLIENT/SERVER` function blocks) and UDP is used for the unidirectional communication model (i.e. `PUBLISH/SUBSCRIBE` function blocks). In order to support these communication protocols, the generic communication function blocks of IEC 61499-1, Annex E.2 (see Section 5.4) are adapted as follows:

- In order to clearly define the number of data inputs and outputs, the type names of the communication SI function blocks are denoted with a numbering scheme:
  - For the bidirectional communication model two numbers in the form `FBTYPE_m_n` are used, where `m` defines the number of inputs and `n` the number of outputs (e.g. `CLIENT_2_3`). If only one number is given, the number of inputs and outputs is the same.
  - For the unidirectional communication model, one number in the form `FBTYPE_m` is used (e.g. `PUBLISH_12`). For the `PUBLISH` function block this is the number of data inputs and for the `SUBSCRIBE` function block the number of data outputs.
  - If one of the numbers given is zero then no data, only an indication for the event called `NULL` data item is transmitted.
- The `PARAMS` input is renamed to `ID` and given the data type `WSTRING`. It specifies the IP[3] address and a port number separated with a ':' to be used for the communication.
  - For the bidirectional communication model, the `SERVER` uses the local IP address of the device and a port number at which it is listening for incoming client connections. The `CLIENT` uses the IP address of the `SERVER`'s device together with the port number specified at the `SERVER`'s `ID` input variable.
  - For the unidirectional communication model, it specifies an IP multicast address (i.e. 224.0.0.1 – 239.255.255.255) and a port number. A PUBLISH/SUBSCRIBE pair need to use the same IP multicast/port number combination.

    Note: Several SUBSCRIBE function blocks can receive the data if they are using the same IP multicast/port number combination as the associated PUBLISH function block.
  - The data type of the output variable `STATUS` is set to `WSTRING` and a set of values for the different statuses of a communication SI function block are defined.

[1] TCP – Transmission Control Protocol.

[2] UDP – User Datagram Protocol.

[3] IP – Internet Protocol.

Finally, it is also defined how data is represented on the network. That means how a set of input variables (i.e. `SD_1, ... SD_m`) is encoded into the payload of a network message so that a receiver can correctly decode it and apply it to its outputs (i.e. `RD_1, ... RD_n`). The format defined is based on the ASN.1 encoding defined in the Annex E.3 of IEC 61499-1. The compliance profile simplifies the encoding of data sequences (i.e. `SD_1, ... SD_m`) to be implicitly encoded one after each other without a sequence header. A `NULL` data element is introduced for communication SI function blocks which have no input variables to send. This allows the transmission of pure event notifications. Furthermore the encoding for derived data types, strings, as well as date and time related data types defined in Annex E.3 of IEC 61499-1 is refined.

B.3 Configurability provisions

The definition for configurability consists of three main parts:

1. Utilisation of a `CLIENT/SERVER` connection between software tools and devices.
2. Refinement of the `MANAGER` function block and its usage in devices.
3. Format for representing the management commands as described in the Sections 2.5 and 5.5.

B.3.1 Configuration communication

The compliance profile defines the use of a `CLIENT/SERVER` connection over Ethernet, as defined in the interoperability provisions, as the communications method to be used for configuration tasks. For this, devices provide a `SERVER`, which is listening for connections at the port number 61499. This allows software tools to connect to devices and perform configuration tasks remotely. If necessary, devices may use a different port number. Therefore, software tools require means that support the configuration of the address and port number of devices.

B.3.2 Configuration interface of a compliant device

The interface to the management functionality of a device is provided in form of a service interface function block, called `DEV_MGR`, which stands for device manager. The `DEV_MGR` function block is roughly based on the `MANAGER` block described in Section 5.3 and is a requester service interface function block type (see Figure B.1). It provides the means by which users of this function block can request the execution of management commands and receive the associated response. The main interface elements of the `DEV_MGR` function block are the input variables `DST` and `RQST` and the `RESP` output variable. All three are of the data type `STRING`. These variables have the following semantics:

`DST` – determines the destination of the management request. The value can either be empty, meaning that the request targets the device, or it can contain

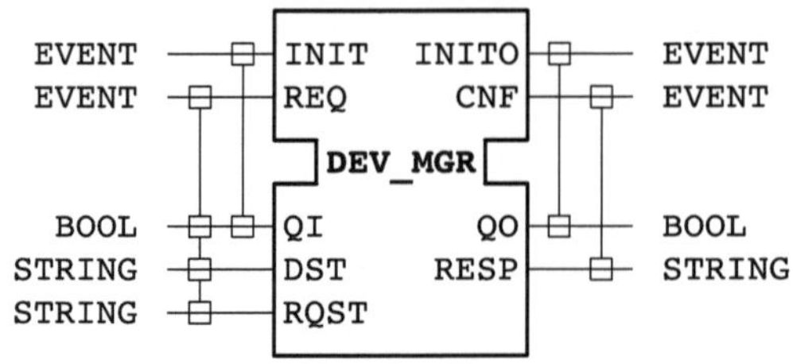

Figure B.1 DEV_MGR function

an identifier of a resource instance, meaning that the resource is the request's target.

RQST – contains the requested command and its data (see below for information on the format used).

RESP– provides information on the success of the command's execution and response data.

Devices can utilise this management functionality in form of the composite function block DM_KRNL as shown in Figure B.2. This composite function block contains a SERVER_1_2 function block connected with a DEV_MGR function block. The SERVER_1_2 function block is in charge of receiving the DST and

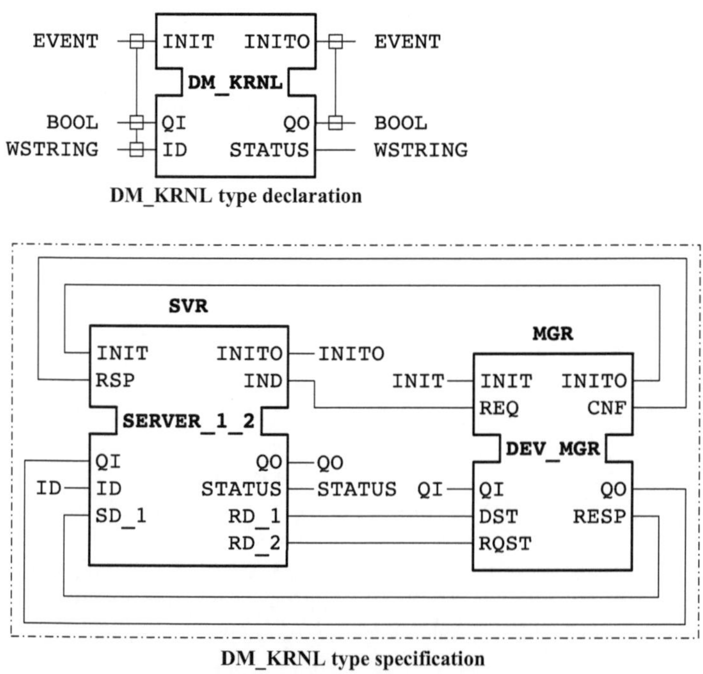

Figure B.2 DM_KRNL function block

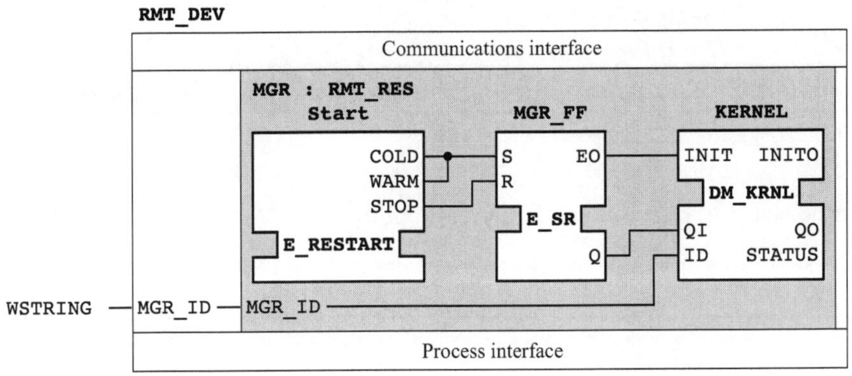

Figure B.3 RMT_DEV device type containing an RMT_RES resource

REQST values from a connected CLIENT, triggering the DEV_MGR and sending the response of the DEV_MGR back to the CLIENT.

In order to reduce the effort of using the DM_KRNL function block by device developers, a resource type with a predefined function block network named MGR_RES is specified. This resource handles the initialisation and de-initialisation of the DM_RKNL function block utilising the E_RESTART and E_SR event function blocks (Figure B.3). Finally, a device type using the MGR_RES is also specified. This device type, named RMT_DEV, has a configuration parameter (MGR) for the address and port number to be used by the SERVER_1_2 (Figure B.3). Using this device type in system configurations allows users to specify the IP address and the port number to be used by the software tool for the configuration connection.

B.3.3 Representation of management commands

IEC 61499 defines only a generic set of nine management commands (see Tables 5.3 and 5.4), which are required to perform the main management tasks of a distributed control system (i.e. download/upload applications). The 'IEC 61499 Compliance Profile for Feasibility Demonstrations' provides a representation format for these commands and their response using an XML representation. The general structure of a command is as follows:

```
< Request ID="..." Action="..." >
   ...   (*Further command data*)
< /Request >
```

The overall command is encapsulated by an XML tag named Request. This tag has two attributes: ID and Action. The value of the ID attribute is a consecutive number which allows software tools to align each request with its corresponding response. The value of the Action attribute defines the requested management command. Possible values for the Action attribute are: CREATE, DELETE, READ, WRITE, START, STOP, KILL, RESET, QUERY. The semantics of these values are according to the descriptions in Tables 5.3 and 5.4.

Table B.1 Management command data

Request data	Data format	Eligible commands
Function block instance	< FB Name="..." Type="..." / >	CREATE, DELETE, START*, STOP*, KILL*, RESET*, QUERY
Connection	< Connection Source="..." Destination="..." / >	CREATE, DELETE, QUERY
Parameter	< Parameter Reference="..." Value="..." / >	READ, WRITE

*These commands can optionally also be used without function block instance data, then it is addressed to the whole device or resource.

Depending on the requested command additional data may be required. This data is also encoded in XML and can be a function block instance, connection or parameter data. Table B.1 shows the structure of this data and the commands which uses it.

Note: Not all attributes are needed in all cases. Readers should consult the compliance profile for more details.

The general structure of a response is as follows:

```
< Response ID="..."Reason="..." >
    ...    (*Further response data*)
< /Response >
```

It uses the XML tag Response with the attributes ID and Reason. The value of the ID is the same as used in the corresponding request. With the Reason value, information on the execution status of the command can be provided. The compliance profile defines a set of Reason values. For example, UNSUPPORTED_TYPE is used to indicate, that in a function block instance request, the requested function block type is not supported by the device. If the command's execution was successful, the Reason attribute is omitted (i.e. < Response ID="..." >). If the response provides additional data, it is encapsulated in a similar form as for the request (Table B.1).

Note: The description given in this appendix is only an overview. It is strongly recommended that software tool and device developers consult the compliance profile specification for further details.

Appendix C
Frequently asked questions (IEC 61499 FAQ)

This appendix is a compendium of frequently asked questions with answers concerning the development and application of the IEC 61499 function block standard. The answers have been provided by international experts working in the IEC working group for IEC 61499 (at that time IEC TC65/WG6). The authors have restructured and added additional comments to some of the answers. Furthermore the authors have updated the comments to be consistent with the current version of the IEC 61499 standard and also provided some additional questions with answers.

> ***Note:*** *This information is not copyrighted and was originally made publicly available and periodically updated at the IEC TC65/WG6 web page.*

C.1 General questions

C.1.1 What are the benefits from using IEC 61499?

IEC 61499-compliant distributed industrial process measurement and control systems (IPMCSs), devices, and their associated life-cycle support systems will be able to deliver a number of significant benefits to their owners and system integrators, including:

1. IEC 61499-compliant life-cycle support systems will be able to **reduce engineering costs** through integrated facilities for configuration, programming and data management. Additional engineering cost savings will result from the ease of system integration provided by IEC 61499's simple yet complete model of distributed systems. This model provides hardware- and operating-system-independent representations for all functions of the system, including control and information processing as well as communications and process interfaces.
2. Economies of scale from uniform application of common software and firmware technologies should **reduce hardware costs**. The most significant cost items in modern control hardware are its embedded firmware and supporting software.
3. Engineers and technicians will be able to **reduce system implementation time** by applying a common set of concepts and skills to all elements of the system. Further reductions in implementation time will result from the elimination of

software patches and 'glueware' formerly required to integrate incompatible system elements and software tools.

4. The elimination of patchware and glueware, the availability of interoperable software toolsets, the portability of engineering skills and the ease of integration of system elements will yield **higher reliability and maintainability** over the system life cycle.
5. IEC 61499 provides an abstract, implementation-independent means of representing system functions. This common 'target' will lead to **simplified migration** from existing systems to IEC 61499-compliant systems, and from older to newer technological platforms (operating systems, communications, etc.) in IEC 61499-compliant systems.

C.1.2 Is IEC 61499 a 'stand-alone' standard?

No, in order to realise the benefits listed in the previous answer, distributed IPMCSs will need:

1. **Communication profiles** which define standard communications function blocks and their mapping to standard open communication services such as those defined in IEC 61158 Fieldbus or IEC 61784.
2. Standard **programming languages** such as those defined in IEC 61131-3 for the specification of algorithms in basic function block types.
3. Standardised **function block types** and guidelines for their application in specific domains, such as the process control function blocks defined in IEC 61805.

C.1.3 Will IEC 61499 have any impact on existing IEC 61131 products?

In the short term there will be no impact to IEC 61131 products. However, there is an intention that the PLC software standard IEC 61131-3 will be revised so that the function block model becomes compatible with IEC 61499. In fact, IEC 61131 function blocks can be easily converted into a form that fits the IEC 61499 model by being encapsulated as basic function blocks. It is also possible to use IEC 61131-3 languages such as Structured Text (ST) and Ladder Diagram (LD) to define algorithms within IEC 61499 function block type specifications.

C.1.4 I already have a huge amount of legacy programs based on IEC 61131-3 and other procedural languages. How can this be reused?

Procedural control algorithms will have to be encapsulated in IEC 61499 basic **function block types**. How quickly and completely this will be possible depends on how soon **software tools** will be available to support this migration. It is certainly fairly straightforward to take algorithms expressed in IEC 61131-3 ST and encapsulate them as IEC 61499 function blocks.

Annex D of IEC 61499-1 describes for this task a light weight IEC 61499 function block type, called simple function block. This function block has just one algorithm and one input and one output event.

There are also first research results available which provide an automatic transformation of a full IEC 61131-3 configuration consisting of resources, programs and function blocks into a functional equivalent IEC 61499 system [25].

C.1.5 Can an application interface with other applications?

No, since there is no application type within which an interface could be defined. However, applications may **communicate** with each other by means of communication function blocks. Also, subapplications may interface with each other via event connections and data connections.

C.1.6 How are applications distributed?

The applications can be distributed using the following process:

1. Create and interconnect one or more instances of the subapplication and function block types representing the application.
2. Create instances of the service interface function blocks representing the process interfaces of the application.
3. Create appropriate event connections and data connections between the process interface function blocks and the subapplications and function blocks representing the application.
4. If necessary to improve the distribution, remove the encapsulation around the subapplications, exposing their component function blocks (which may also be subapplications) as re-distributable elements of the application.
5. Remove the encapsulation of all of the newly exposed subapplications, repeating as necessary.
6. Distribute the application by allocating its function block instances to appropriate resources.
7. Create and configure appropriate communication function block instances within the resources to maintain the event and data flows of the application.

Note: It is envisaged the engineering support tools will automatically insert service interface block blocks to maintain event and data flows between connected blocks when they are distributed to run on different resources.

C.1.7 Why can't a composite function block have internal variables?

Internal variables are not required in composite function blocks, since sampling and storage is defined for all possible sources of data, i.e. input or outputs variables of component function block instances.

C.1.8 What is the difference between Composite Function Block and Subapplication?

Composite function block has guarantees for sampling of input and output data when these are associated with events. For this reason, it is not distributable, i.e. it is not possible to break the internal connections between component blocks and run them on different processing resources. However, to achieve the same effect, this can be done by converting the composite block type into a subapplication type.

C.1.9 Can alternative graphical representations be used?

Software tools can use alternative graphical, textual or tabular representations, so long as accompanying documentation specifies an unambiguous mapping to the graphical elements and associated textual syntax defined in IEC 61499-1. It should be noted that IEC 61499 defines a model not a graphical programming language.

C.1.10 How can a function block respond to faults and exceptions?

In the IEC 61499 model, faults and exceptions are modelled as event outputs and associated data outputs of function blocks. These outputs can be connected to appropriate event inputs and associated data inputs of any function blocks, which must respond to the fault or exception, e.g. by changing their operational **modes**. (*Modes are discussed in Chapter 7, conditional event propagation for triggering fault handling chains are discussed in Chapter 6.*)

C.1.11 How is the loading and starting of applications managed?

Software tools for this purpose will take as input a system configuration and generate sequences of commands to management function blocks to perform loading and starting of applications, either locally or remotely via communication function blocks.

Note: The communications protocol and interactions to load and start-up applications are not defined within IEC 61499 and are not part of its scope. This should be defined in compliance profiles (an example of such a compliance profile is given in Appendix B). However, IEC 61499 does define a textual syntax and file exchange format to allow applications to be ported between systems.

C.1.12 Why are there no GLOBAL or EXTERNAL variables?

All variables are encapsulated. There is no guarantee that there will be an implicit global distribution mechanism available. When such mechanisms are available, they can always be mapped to service interface function blocks. For example, configuration parameters can be encapsulated in a function block that is then connected to other blocks that require the values.

C.1.13 What is an ECC? Why should it be used and when?

An execution control chart (ECC) is a specialised state machine to enable multiple events to trigger multiple algorithms, possibly dependent on some internal state. It should be used when it is necessary to have maximum flexibility in algorithm selection and scheduling, or when there is a requirement for a high-performance, event-driven state machine. It is defined in IEC 61499 as a formal and precise way to show how events can trigger the execution of internal algorithms. (*This is discussed in Chapter 3.*)

C.1.14 Why use function blocks to model device or resource management applications?

This usage is described in subclauses 4.7, 6.3 and Annex E of IEC 61499-1. The benefits of using function blocks to model device and resource management are:

1. A consistent model of all applications in the system, including management applications.
2. A consistent means of encapsulating and reusing all functions, including management functions.
3. Reuse of existing data types.
4. Use of existing, standardised means for the definition of required new data types and specification of management messages.

C.2 Object orientation questions

C.2.1 Why use such a heavily object-oriented model?

The degree of object orientation used in IEC 61499 is necessary in order to achieve the required level of distributability of encapsulated, reusable software modules (function blocks).

Note: There are limitations to the OO capabilities provided in IEC 61499, e.g. IEC 61499 does not support multiple-inheritance.

C.2.2 Isn't IEC 61499 model very expensive to implement?

Not necessarily; a full object-oriented **implementation** is not required by the IEC 61499 **model** – only that the externally visible behaviours of a compliant **device** conform to the requirements for the device compliance class as defined in Annex B of IEC 61499-2. The implementation of a 'Class 0' simple device can be quite economical. (*Compliance classes are discussed in Chapter 8.*)

C.2.3 Why not use a general-purpose distributed object model, like DCOM or CORBA?

There are a number of reasons:

1. Implementation of the features specified for these information-technology models would be too expensive, and their performance would almost always be too slow, for use in a distributed real-time IPMCS.
2. There is no standard, easily understood graphical model for representing the interconnections of events and data among these kinds of objects in distributed applications.
3. The function block paradigm is closer to the way system and process engineers think about control functionality.

C.2.4 Are data and event connections a kind of object?

Data and event connections can be considered objects insofar as IEC 61499 defines a textual syntax for their declaration and a management service for their creation, query and deletion. However, the only attributes that IEC 61499 defines for a connection are its source and destination. Software tools may require additional attributes, such as scheduling priority for event connections, communication timeliness constraints for both data and event connections in order to determine whether a proposed system configuration is feasible. Software tools should also check the types of data and event inputs and outputs for consistency. Annex G of IEC 61499-1 defines syntax and semantics for the declaration of additional attributes.

C.3 The event-driven model questions

C.3.1 What is the difference between event and data?

More formally the difference can be expressed as:

- **Event:** An instantaneous occurrence that is significant to scheduling the execution of an algorithm.
- **Data:** A reinterpretable representation of information in a formalised manner suitable for communication, interpretation or processing.

It is important to note that an event signifies some form of state change that may signal that an algorithm should be run. In practical terms, it is sometimes difficult to decide when an event is required or whether a state change can be reflected in the value of an input or output. For example, a function block mode change could be signalled by a dedicated mode input event. Alternatively, it could also be signalled by a mode input variable and an update event.

Also note that data cannot be transferred between blocks without an accompanying event.

C.3.2 Why use an event-driven model?

Any execution control strategy (cyclic, time scheduled, etc.) can be represented in terms of an event-driven model, but the converse is not necessarily true. IEC 61499 opts for the more general model in order to provide maximum flexibility and descriptive power to compliant standards and systems.

C.3.3 How can a change in a data value generate an event?

E_R_TRIG and E_F_TRIG function block types are defined in Annex A of IEC 61499-1 which propagate an input event when the value of a boolean input rises or falls, respectively, between successive occurrences of the input event. Instances of these types may be combined with other function blocks to produce rising- and falling-edge triggers, threshold detection events, etc. (*A discussion of standard event function blocks is given in Chapter 6.*)

C.3.4 What are event types? What are they used for?

An *event type* is an identifier associated with an event input or an event output of a function block type, assigned as part of the event input or output declaration, as described in subclause 5.2.1.2 of IEC 61499-1. It can be used by software tools to assure that event connections are not improperly mixed; for instance to ensure that an event output that is intended to be used for initialisation is not connected to an event input that is intended to be used for alarm processing.

Event types cannot be detected by ECCs, so the type of an event cannot be used to influence the processing of events in basic function block types.

Note: IEC 61499 does not define any standard event types other than the default type `EVENT`. Domain standards may define their own specific event types.

C.3.5 Why and when should the WITH construct be used?

This is used when designing function block types; the WITH construct is used to specify:

1. which input variables to sample when an event occurs at the associated event input of an instance of the type;
2. which output variables to sample when an event occurs at the associated event output of an instance of the type.

When using the WITH construct, it should be noted that this information will be used by software tools to assist the designer of applications using instances of these types to ensure that:

1. the data used by an algorithm in one function block is consistent with the data produced by an algorithm in another function block and delivered over one or more data connections associated with one or more event connections;
2. the messages transmitted over communication connections between resources in a distributed application carry consistent data and events between the function block instances in the application in the way intended by the application designer.

C.3.6 How can this model accommodate sampled-data systems?

The major issues in sampled-data systems such as those used in motion, robotic and continuous process control are:

1. how to achieve synchronised sampling of inputs;
2. how to ensure that all data has arrived in time for processing by control algorithms;
3. how to ensure that all outputs are present and ready for sampling before beginning the next cycle of sampling and execution.

Solutions to these problems typically require specialised communication and operating system services, which can be represented in terms of the IEC 61499 model by service interface function blocks. The inputs and outputs to be sampled can likewise be represented by service interface function blocks, while the algorithms to be performed will typically be encapsulated in basic function blocks. The relationships between the system services, input and output sampling, and algorithm execution can then be expressed as event connections and data connections.

C.3.7 What happens if a critical event is lost by the communication sub-system?

This issue is common to all distributed control systems and its solutions are well known, for example, via periodic communication, missing event detection and/or positive acknowledgement protocols. The IEC 61499 model provides for notification of abnormal operation via the `IND-`, `CNF-` and `INITO-` service primitives and `STATUS` output of service interface function blocks (see Chapter 5).

C.4 Engineering methodologies

C.4.1 How can I use IEC 61499 to design and implement state-machine control?

There are various formal and informal methodologies for the design of state-machine control systems. A general outline of these methods and how IEC 61499 models and software tools can be used is as follows:

1. Define the desired **sequence of operations** for the controlled machine or process, as well as possible abnormal sequences that may occur. This may be done informally in ordinary language, or using more formal notations such as Petri nets or IEC 61131-3 Sequential Function Charts (SFCs).
2. Define the **actuators** that are to be used to implement the desired behaviours, the **sensors** that are to be used to determine the actual state of the controlled machine or process, and their **interfaces** to the state-machine controller(s). IEC 61499 service interface function block types may be used for this purpose; such type definitions will typically be provided for this purpose by the supplier of the IEC 61499-compliant sensor and actuator devices.
3. **Model** the behaviours of the machine or process in response to **commands** (events plus data) given to the **actuators**, and the resulting, observable **time-dependent outputs** (events plus data) from the **sensors**. This modelling may be done formally using notations such as Petri nets, or informally using simulation models (which may be implemented with appropriate IEC 61499 models).
4. Define the appropriate **state-machine controllers**, typically as the ECCs and algorithms of basic function block types. Methodologies for doing this step may be informal, based on engineering experience, or more formal, using for example Petri-net theory. Best of all is to **reuse** existing, proven state-machine controllers from an IEC 61499 function block library!

5. **Validate** the proposed state-machine controllers versus the **model** of the controlled machine or process. In order to catch possible design or specification errors, this should normally be done using **simulation** in addition to more formal validation methods if available.
6. Finally, replace the **simulated sensor and actuator interfaces** with the **service interfaces** to the actual machine or process to be controlled. This installation is normally to be done piecewise for testing purposes.

Appendix D
PID function block example

This appendix depicts how a Proportional, Integral and Derivative (PID) algorithm can be encapsulated as a composite function block.

A PID algorithm is used in 'closed loop' control where there is a requirement to control a process that may be subject to disturbances caused by external factors or by unpredictable changes to the process. Examples are: to control the temperature of a heat treatment oven where the internal temperature has to remain stable while the oven door is opened and different loads are inserted into the oven, or to control the pressure in a reactor vessel while a chemical reaction is progressing. The PID algorithm uses the error between the current Process Value (PV) and the desired process value or Set-Point (SP), along with the integral and derivative values of the error, to calculate a new output value to drive the process. This output value is used to drive an actuator to, say, adjust the current in a heater or change a pump speed in order to bring the process value closer to the SP. A PID algorithm has a number of parameters that must be 'tuned' to match the process under control. This is required to ensure that the control action is responsive and has minimum overshoot. An application using a PID is discussed in Section 7.1. The main features of the PID algorithm are shown in Figure D.1.

Note: This example is intended to demonstrate the use of IEC 61499 to encapsulate algorithms. The design of PID algorithms is a complex subject and

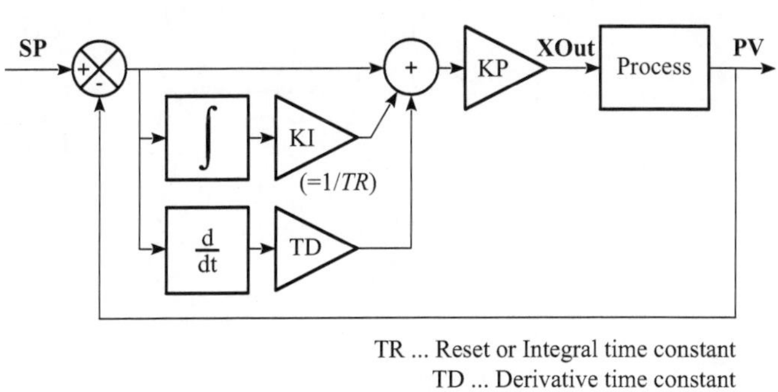

Figure D.1 PID algorithm

outside the scope of this book. An industrially robust PID function block would require significantly more functionality than is described in this appendix.

D.1 PID function block specification

In this example, the PID function block is constructed by connecting instances of three function blocks, `INT1`, `DEV1` and `CALC1`. These are created from `DerivativeReal`, `IntegralReal` and `PID_Calc` function block types respectively. The `DerivativeReal` and `IntegralReal` function blocks provide derivative and integration functionality for inputs of `REAL` data type and are based on definitions given as examples in the IEC 61131-3 PLC software

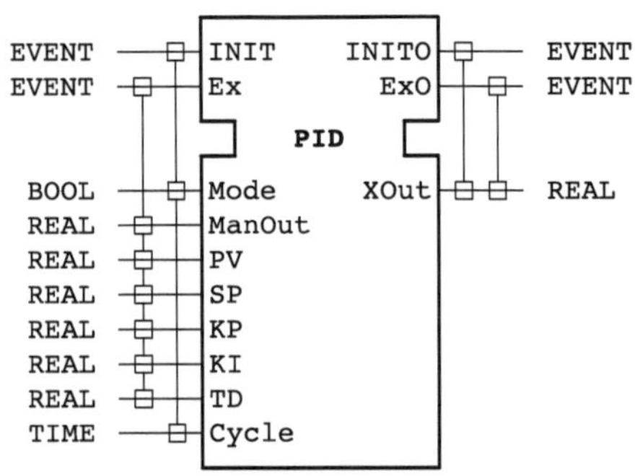

PID type declaration

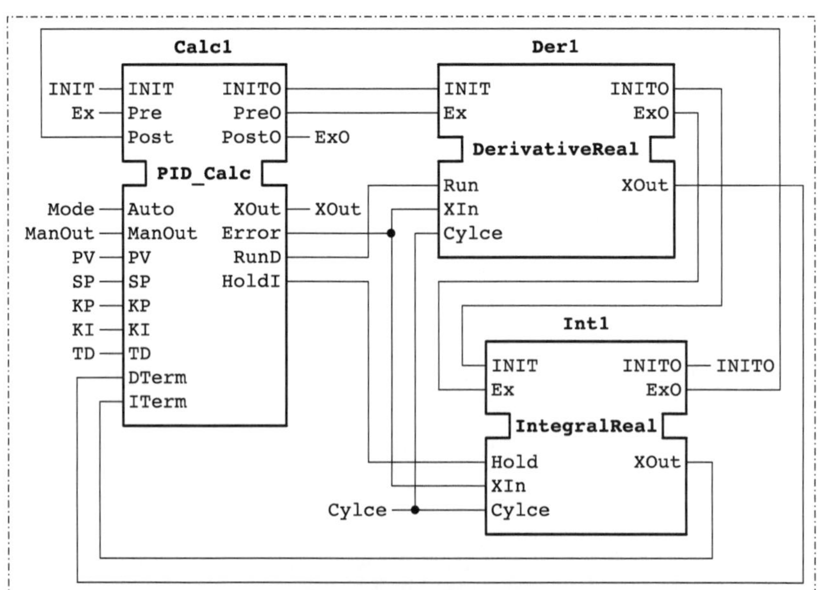

PID function block type specification

Figure D.2 PID function block type

standard. The `PID_Calc` block encapsulates the PID algorithm and calculates a new output value that is proportional to a weighed summation of the error, the derivative of the error and the integral of the error (Figure D.2).

The `PID_Calc` function block type has two execution states `Pre` and `Post`. During the `Pre` state, it runs an algorithm to calculate the error between SP and PV. It then generates an output event `PreO` that is used to trigger the execution of the derivative and integral function blocks that are external to this block. These external blocks use the error value calculated by the Alg`Pre` algorithm. The `Post` state is activated by input event `Post`, and is entered when the external derivative and integral blocks have completed their execution. The Alg`Post` algorithm uses the values of error derivative and error integral from the external blocks, along with the current error to calculate the new value for the output `XOut`.

The `PID_Calc` block has two modes, 'AUTO' and 'MANUAL' set by the input `Auto` when it is set to `TRUE` and `FALSE`, respectively. In the AUTO mode, the PID algorithm is used to update the output `XOut`. When the `Auto` input is `FALSE`, the PID algorithm is not active and the block is set to 'MANUAL' mode. In this mode, the output `XOut` is set directly by a manually supplied value given by input `ManOut`.

D.1.1 Textual syntax of PID function block

```
(* PID implementation as a composite   *)
(* function block. This is constructed *)
(* using the IntegralReal,             *)
(* DerivativeReal, and PID_Calc FBs    *)
FUNCTION_BLOCK PID
  EVENT_INPUT
    INIT WITH MODE, Cycle;
    Ex WITH ManOut, PV, SP, KP, KI, TD;
  END_EVENT
  EVENT_OUTPUT
    INITO WITH XOut;
    ExO WITH XOut;
  END_EVENT
  VAR_INPUT
    MODE : BOOL;    (* 1=Auto or 0=Manual *)
    ManOut : REAL; (* Manual set output *)
    PV : REAL;      (* Process value *)
    SP : REAL;      (* Setpoint *)
    KP : REAL;      (* Proportional gain *)
    KI : REAL;      (* Integration constant 1/sec *)
    TD : REAL;      (* Derivative time, sec *)
    Cycle : TIME; .(* Sampling period *)
  END_VAR
  VAR_OUTPUT
     XOut : REAL;   (* PID output *)
```

```
  END_VAR
  (* Function blocks *)
  FBS
    Int1 :  IntegralReal;
    Der1 :  DerivativeReal;
    Calc1 : PID_Calc;
  END_FBS
  EVENT_CONNECTIONS
    INIT TO Calc1.INIT;
    Calc1.INITO TO Der1.INIT;
    Der1.INITO TO  Int1.INIT;
    Int1.INIT TO INITO;
    Ex TO Calc1.Pre;
    Calc1.PreO TO Der1.Ex;
    Der1.ExO TO Int1.Ex;
    Int1.ExO TO Calc1.Post;
    Calc1.PostO TO ExO;
  END_CONNECTIONS
  DATA_CONNECTIONS
    Mode TO Calc1.Auto;
    PV TO Calc1.PV;
    SP TO Calc1.SP;
    KP TO Calc1.KP;
    KI TO Calc1.KI;
    TD TO Calc1.TD;
    ManOut TO Calc1.ManOut;
    Calc1.Error TO Int1.XIn;
    Calc1.Error TO Der1.XIn;
    Calc1.RunD TO Der1.Run;
    Calc1.HoldI TO Int1.Hold;
    Der1.XOut TO Calc1.DTerm;
    Int1.XOut TO Calc1.ITerm;
    Cycle TO Der1.Cycle;
    Cycle TO Int1.Cycle;
    Calc1.XOut TO XOut;
  END_CONNECTIONS
END_FUNCTION_BLOCK
```

D.2 PID component function blocks

The graphical function block type declarations required to create the PID are depicted in Figures D.3–D.5. The IEC 61499 Textual Syntax that describes the specification for each of the blocks is also given. These specifications include algorithms expressed using the IEC 61131-3 Structured Text language. At the end of this appendix is an example of an algorithm expressed using the Java language.

D.2.1 DerivativeReal function block specification

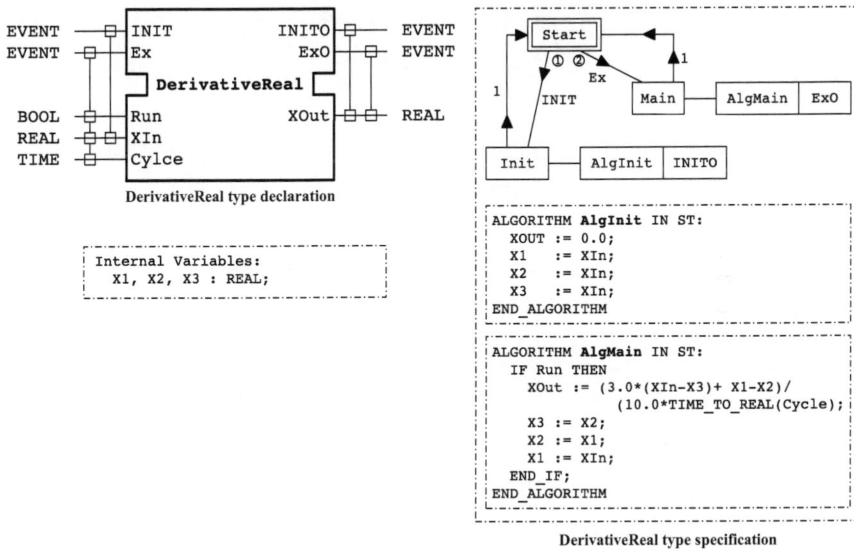

Figure D.3 DerivativeReal function block type

```
(* DerivativeReal *)
(* This function block produces an output XOut *)
(* proportional to the rate of change of the   *)
(* input XIn. The derivative is calculated     *)
(* while the Run input is true. The Cycle time *)
(* is required to calculate the change of      *)
(* input value over time.                      *)
FUNCTION_BLOCK DerivativeReal
  EVENT_INPUT
    INIT WITH XIn;
    Ex WITH Run, XIn, Cycle;
  END_EVENT
  EVENT_OUTPUT
    INITO WITH XOut;
    ExO WITH XOut;
  END_EVENT
  VAR_INPUT
```

```
    Run : BOOL;    (* 1 = Run 0 = Hold *)
    XIn : REAL;    (* Derivative input *)
    Cycle : TIME; (* Scan cycle time *)
  END_VAR
  VAR_OUTPUT
    XOut : REAL;   (* Derivative output *)
  END_VAR
  VAR
    X1, X2, X3 : REAL;
  END_VAR
  EC_STATES
    Start;
    Init:    AlgInit -> INITO; (* Initial state *)
    Main:    AlgMain -> ExO;   (* Main state *)
  END_STATES
  EC_TRANSITIONS
    Start TO Init  := INIT;
    Start TO Main  := Ex;
    Init TO  Start := 1;
    Main TO  Start := 1;
  END_TRANSITIONS
  (* Algorithms expressed using Structured Text *)
  ALGORITHM AlgInit IN ST: (* Initialisation *)
    XOut := 0.0;
    X1   := XIn;
    X2   := XIn;
    X3   := XIn;
  END_ALGORITHM
  ALGORITHM AlgMain IN ST: (* Main algorithm *)
    IF Run THEN
      XOut := (3.0 * (XIn-X3) + X1 - X2)/
                  (10.0 * TIME_TO_REAL(Cycle));
      X3 := X2;
      X2 := X1;
      X1 := XIn;
    END_IF;
  END_ALGORITHM
END_FUNCTION_BLOCK
```

D.2.2 IntegralReal function block specification

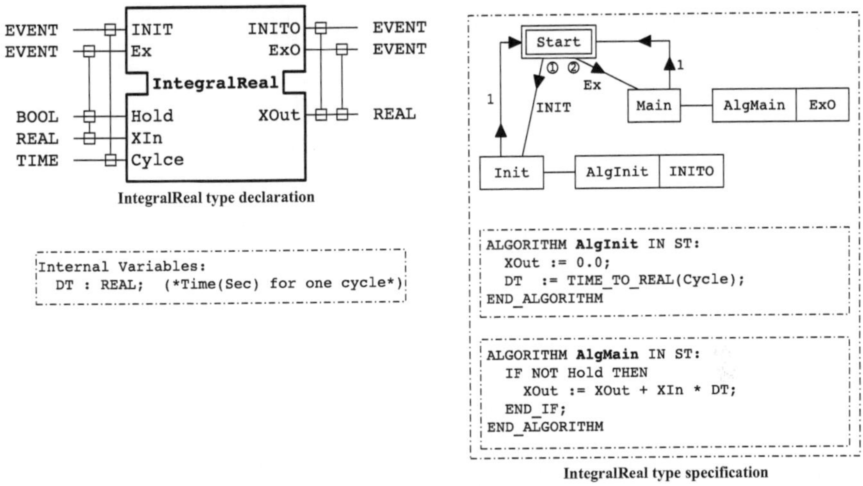

Figure D.4 IntegralReal function block type

```
(* IntegralReal *)
(* This function block integrates the value of *)
(* input XIn over time. The integration can be *)
(* reset by an initialisation event INIT. The  *)
(* Cycle time defines the time between block   *)
(* update events Ex. The integration continues *)
(* while the Hold input is false. If it is     *)
(* true, the integration value is frozen.      *)
FUNCTION_BLOCK IntegralReal
  EVENT_INPUT
    INIT WITH Cycle;
    Ex WITH Hold, XI;
  END_EVENT
  EVENT_OUTPUT
    INITO WITH XOut;
    ExO WITH XOut;
  END_EVENT
  VAR_INPUT
    Hold : BOOL;    (* 0 = Run, 1 = Hold *)
    XIn : REAL;     (* Integrand *)
    Cycle : TIME;   (* Scan cycle time *)
  END_VAR
  VAR_OUTPUT
    XOut : REAL;    (* Integrated output *)
  END_VAR
```

```
  VAR
    DT : REAL;      (* Time(Sec) for one cycle *)
  END_VAR
  EC_STATES
    Start;
    Init:    AlgInit -> INITO; (* Initial state *)
    Main:    Main -> ExO;      (* Main state *)
  END_STATES
  EC_TRANSITIONS
    Start TO Init := INIT;
    Start TO Main := Ex;
    Init TO Start := 1;
    Main TO Start := 1;
  END_TRANSITIONS
  (* Algorithms expressed using Structured Text *)
  ALGORITHM AlgInit IN ST: (* Initialisation *)
    XOut := 0.0;
    DT  := TIME_TO_REAL(Cycle);
  END_ALGORITHM
  ALGORITHM MAIN IN ST:    (* Main algorithm *)
    IF NOT Hold THEN
      XOut := XOut + XIn * DT;
    END_IF;
  END_ALGORITHM
END_FUNCTION_BLOCK
```

D.2.3 PID_Calc function block specification

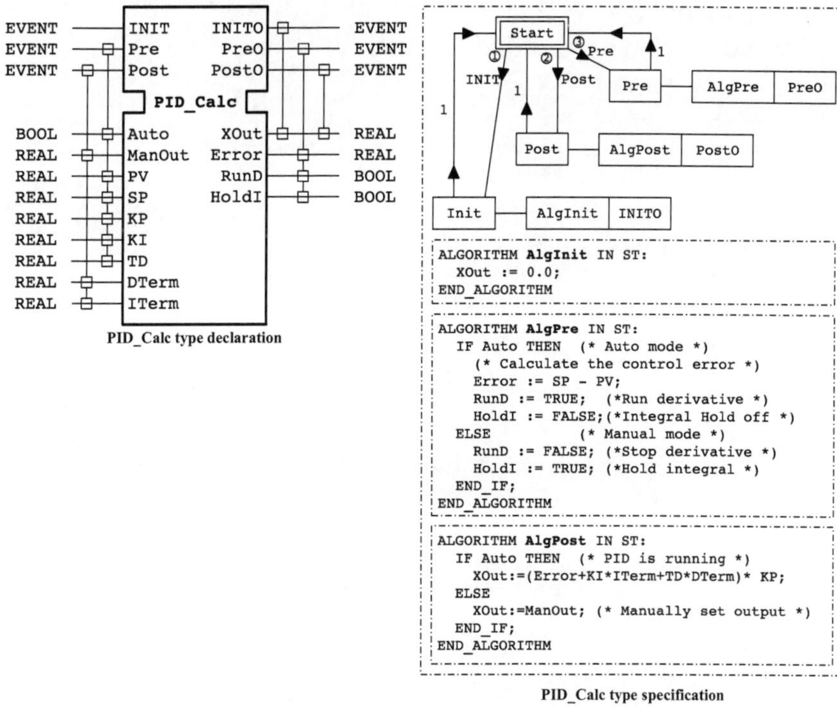

Figure D.5 PID_Calc function block type

```
(* PID_Calc *)
(* The function block provides the              *)
(* calculations to create the PID algorithm.    *)
(* It has two main states Pre and Post. During *)
(* the Pre state the error between the         *)
(* set-point (SP) and process value (PV) is    *)
(* calculated. During the Post state, the      *)
(* value for the PID output XOut is            *)
(* calculated using the current error,         *)
(* integrated error and derivative error.      *)
(* Auto input when true allows the PID         *)
(* to run, when false, sets output to the      *)
(* manual output value, i.e. MANUAL mode.      *)
FUNCTION_BLOCK PID_Calc
  EVENT_INPUT
    INIT;
    Pre WITH Auto, PV, SP, KP, KI, TD;
    Post WITH Manout, ITerm, DTerm;
  END_EVENT
```

```
EVENT_OUTPUT
  INITO WITH XOut;
  PreO WITH Error;
  PostO WITH RunD, HoldI, XOut;
END_EVENT
VAR_INPUT
  Auto : BOOL;    (* 1 = Auto 0 = Manual *)
  ManOut : REAL;  (* Manual output value *)
  PV : REAL;      (* Process value *)
  SP : REAL;      (* Setpoint *)
  KP : REAL;      (* Proportionality or(gain) *)
  KI  : REAL;     (* Integration constant 1/sec *)
  TD : REAL;      (* Derivative time, sec *)
  ITerm : REAL;   (* Integral of error *)
  DTerm : REAL;   (* Derivative of error *)
END_VAR
VAR_OUTPUT
  XOut : REAL;    (* PID output *)
  Error : REAL;   (* Error output *)
  RunD : BOOL;    (* Run the derivative *)
  HoldI : BOOL;   (* Hold the integral *)
END_VAR
EC_STATES
  Start;
  Init:  AlgInit -> INITO; (* Initial state *)
  Pre:   AlgPre -> PreO;   (* Pre state *)
  Post:  AlgPost-> PostO;  (* Post state *)
END_STATES
EC_TRANSITIONS
  Start TO Init := INIT;
  Init TO Start := 1;
  Start TO Pre := Pre;
  Pre TO Start := 1;
  Start TO Post := Post;
  Post TO Start := 1;
END_TRANSITIONS
(* Algorithms expressed using Structured Text *)
ALGORITHM AlgInit IN ST: (* Initialisation *)
  XOut := 0.0;
END_ALGORITHM
ALGORITHM AlgPre IN ST: (* Pre algorithm *)
  IF Aute THEN
    (* Auto mode *)
    (* Calculate the control error *)
    Error := SP - PV;
```

```
      RunD := TRUE;    (* Run derivative *)
      HoldI := FALSE; (* Integral Hold off *)
    ELSE
      (* Manual mode *)
      RunD := FALSE; (* Stop derivative *)
      HoldI := TRUE; (* Hold integral *)
    END_IF;
  END_ALGORITHM
  ALGORITHM AlgPost IN ST: (* POST algorithm *)
    (* Calculate the new output value *)
    IF AUTO THEN
      (* PID is running *)
      XOut:=(Error + KI * ITerm + TD * DTerm) * KP;
    ELSE
      XOut := ManOut; (* Manually set output *)
    END_IF;
  END_ALGORITHM
END_FUNCTION_BLOCK
```

D.3 Algorithms expressed using Java

The IEC 61499 standard allows algorithms to be expressed in alternative languages provided there is a clear and unambiguous mapping between the language variable data types and the IEC data types. In this PID example, the algorithms defined in the Textual Syntax for `DerivativeReal`, `IntegralReal` and `PID_Calc` could be expressed in other languages such as Java or C.

To illustrate this, the `AlgInit` and `AlgMain` algorithms for the `DerivativeReal` function block could be expressed in Java as follows:

```
(* Derivative_real algorithms *)
(* Algorithms expressed using Java *)
ALGORITHM AlgInit IN Java: (* Initialisation *)
  {
    Xout.value = 0.0;
    X1.value   = XIn.value;
    X2.value   = XIn.value;
    X3.value   = XIn.value;
  }
END_ALGORITHM
ALGORITHM AlgMain IN Java: (* Main algorithm *)
  {
    if ( Run.value ) {
      // Calculate derivative value
      XOut.value = (3.0 * (XIn.value - X3.value) +
                    X1.value - X2.value) /
                   (10.0 *(real)Cycle.value);
```

```
        X3.value = X2.value;
        X2.value = X1.value;
        X1.value = XIn.value;
      }
    }
END_ALGORITHM
```

D.4 Testing function block models using a software engineering tool

A major benefit that comes from developing function blocks using IEC 61499, due to using a standard methodology, is that it opens up the possibility for software tools to be developed for editing, parsing and testing function blocks. At the time of writing this edition several software tools for IEC 61499 development were available. These tools allow the development of function block specifications of all function block types graphically or using textual syntax. The function blocks can then be used in applications, subapplications and device specific configurations. Different testing facilities allow the evaluation of the behaviour of the developed function block types and applications. For this investigation, we will consider the following three IEC 61499 compliant engineering tools:

FDBK which stands for Function Block Development Kit has been developed by Dr. Christensen from Holobloc Inc. FBDK is the first IEC 61499 software tool and helped a lot in evaluating and refining the concepts of the standard. FBDK is freely available for non-commercial use at: http://www.holobloc.com.

nxtSTUDIO is a commercial IEC 61499 software tool from the company nxtControl. Apart from using IEC 61499 for modelling distributed industrial process measurement and control systems, it also provides tight integration with Human–Machine Interface (HMI) development reducing the effort of connecting the control system with the HMI. Further information on nxtSTUDIO is available at: http://www.nxtcontrol.com/en.

4DIAC-IDE is the IEC 61499 software tool developed within the open source project 4DIAC, which stands for Framework for Distributed Industrial Automation and Control. 4DIAC provides a complete open source implementation of IEC 61499 and was the first fully event-driven implementation providing real-time constrained execution of IEC 61499. Further information can be found at: http://www.fordiac.org.

These tools store the function block types according to the compliance provision of the 'IEC 61499 Compliance Profile for Feasibility Demonstrations' (see Appendix B) in the XML format defined in IEC 61499-2. This allows function block types developed in one of the tools to be reused in any of the others. Figure D.6 shows this with screenshots of the interface and the ECC of the `PID_Calc` function block, opened in all three tools.

These tools provide means by which the user can test a particular block by defining values in a test window for each input and then running the internal

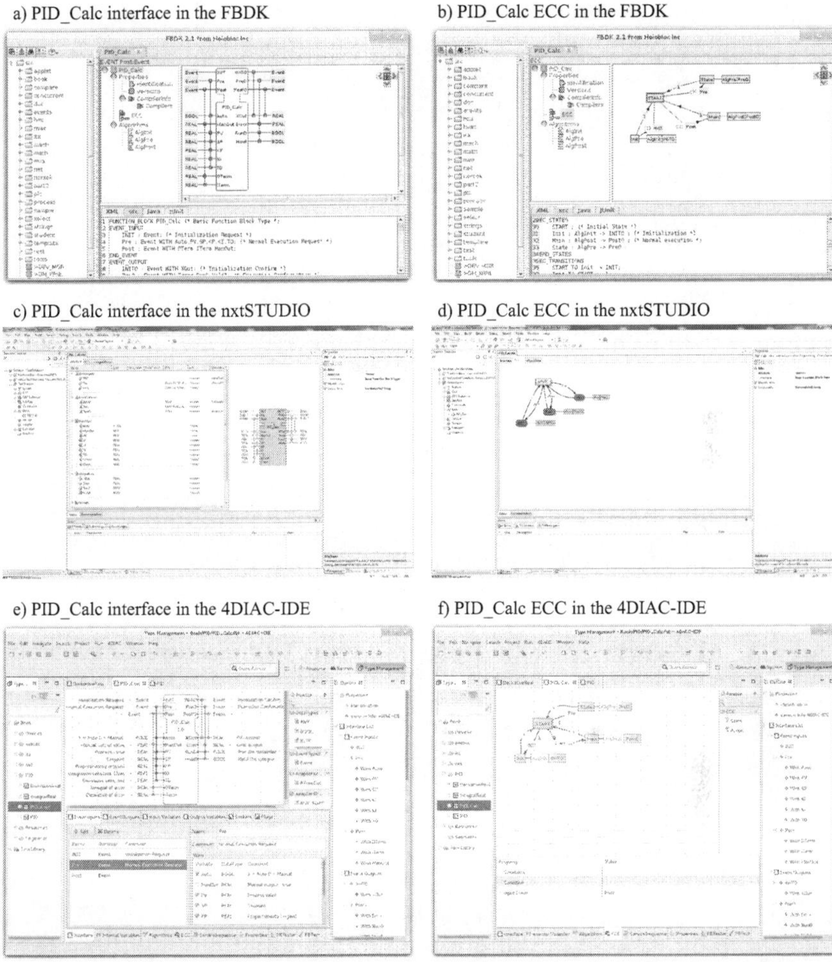

Figure D.6 PID_Calc function block interface and ECC shown in the different software tools

algorithm. The values of the block outputs are updated each time the algorithm is executed when triggered by the user. The user can then at any time compare the output values displayed in the window with those expected. The screenshot in Figure D.7 shows the `DerivativeReal` function block while it is being tested using the 4DIAC-IDE and depicts the test window with input and output values after the algorithm has been executed.

Note: Blocks such as the `DerivativeReal` function block have internal variables so that it may be necessary to run the algorithm many times with the same input values to check that it iterates to the correct output values.

Figure D.8 depicts a further powerful feature of IEC 61499 when used with the associated software tools; it allows the representation of plant behaviour in function blocks. This facilitates the testing of applications against a simulated plant – so

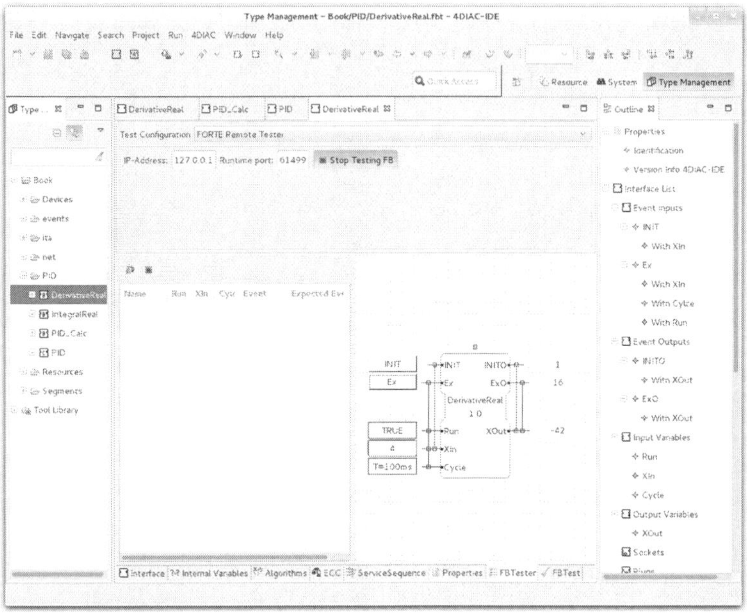

Figure D.7 Screenshot of testing the DerivativeReal function block in the 4DIAC-IDE

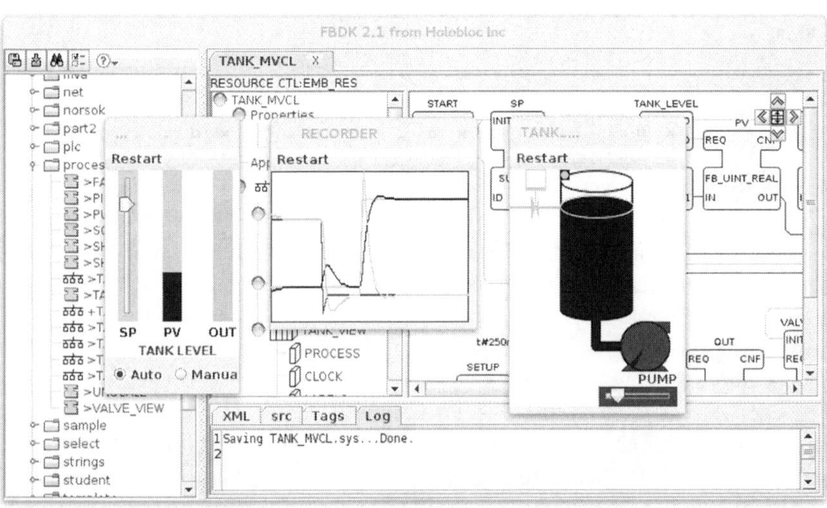

Figure D.8 Testing the PID function block with a simulated tank in the FBDK

enabling applications to be evaluated during early stages in the design. In this screenshot, we see the FBDK, where the `PID` function block is being used in a simple application to control the fluid level in a tank.

Note that the screenshots depicted in:

- Figures D.6a, D.6b and D.8 were produced using the FBDK tool developed by Dr. J. H. Christensen, and are shown courtesy of Holobloc Inc.
- Figures D.6c and D.6d were produced using the nxtStudio tool and are shown courtesy of nxtControl GmbH.
- Figures D.6e, D.6f, and D.7 were produced using the open source tool 4DIAC-IDE.

Appendix E
Exchange formats

This appendix gives a brief overview of the main keywords and structure used in the IEC 61499 Textual Syntax and XML format. It is provided to assist with the understanding of the examples of Textual Syntax and XML format given in this book. For a full description and the full and formal production rules of the Textual Syntax, the reader is advised to read Annex B of IEC 61499-1. For a full description of the XML format and the associated Document Type Definition, the reader is advised to read Annex A of IEC 61499-2.

Text enclosed in quotes '. . . ' are keywords; normal text are identifiers (e.g. function block type name); text enclosed in comments (. . . *) is descriptive.*

E.1 Function block type specification elements

The elements for specifying a function block type are the function block type definition itself containing the static and dynamic interface specification. The static interface consists of the event and data inputs and outputs, and of the adapter plugs and sockets. The dynamic interface is defined with the service declaration containing service sequences.

E.1.1 Function block type definition

SYNTAX:

```
'FUNCTION_BLOCK' FBTypeName         <FBType Name=" FBTypeName">
     (*function block type specification body*)
'END_FUNCTION_BLOCK'                </FBType>
```

This is used to define a function block type specification. The specification body must contain the function block interface list and can contain, where appropriate, a service declaration. This identifies the input and output events and data passed across the interface into and out from the function block algorithms (for basic blocks) or component function blocks (for composite blocks).

For a basic function block, the specification body also contains internal variables, the execution control chart declaration and the algorithm declarations.

For composite function blocks, the specification contains a list of component function block instances, and a list of connections between the components.

E.1.2 Event input/output specification

SYNTAX:

```
'EVENT_INPUT'                    <EventInputs>
  (*event input declarations*)
'END_EVENT'                      </EventInputs>
'EVENT_OUTPUT'                   </EventOutputs>
  (*event output declaration*)
'END_EVENT'                      </EventOutputs>
```

These keywords are used to define the input and output events in a function block specification body, i.e. it defines events that pass into and out of the function block interface. Each event, input and output declaration, contains the event name and an optional event type. The 'WITH' keyword can be used to associate an event with one or more data inputs (for input events) or data outputs (for output events).

For example,

```
EVENT_INPUT
  INIT : INIT_EVENT;
  RUN WITH MODE,PV;
END_EVENT
EVENT_OUTPUT
  INITO : INIT_EVENT WITH XOUT;
  EXO WITH XOUT;
END_EVENT

<EventInputs>
  <Event Name="INIT" Type="INIT_EVENT"/>
  <Event Name="RUN">
    <With Var="MODE"/>
    <With Var="PV"/>
  </Event>
</EventInputs>
<EventOutputs>
  <Event Name="INITO" Type="INIT_EVENT">
    <With Var="XOUT"/>
  </Event>
  <Event Name="EXO" Type="INIT_EVENT">
    <With Var="XOUT"/>
  </Event>
</EventOutputs>
```

E.1.3 Data input/output specification

SYNTAX:

```
'VAR_INPUT'                  <InputVars>
  (*input variable declarations*)
```

```
'END_VAR'                    </InputVars>
'VAR_OUTPUT'                 <OutputVars>
  (*output variable declarations*)
'END_VAR'                    </OutputVars>
```

These keywords are used to define the input and output variables in a function block specification. Each input and output declaration contains the variable name and its data type.

For example,

```
VAR_INPUT
  MODE : BOOL;
  PV : REAL;
  CYCLE : TIME;
END_VAR
VAR_OUTPUT
  XOUT : REAL;
END_VAR

<InputVars>
  <VarDeclaration Name="MODE" Type="BOOL"/>
  <VarDeclaration Name="PV" Type="REAL"/>
  <VarDeclaration Name="CYCLE" Type="TIME"/>
</InputVars>
<OutputVars>
  <VarDeclaration Name="XOUT" Type="REAL"/>
</OutputVars>
```

E.1.4 Adapter plug list specification

SYNTAX:

```
'PLUG'                       <Plugs>
  (*plug declarations*)
'END_PLUGS'                  </Plugs>
```

This defines a list of adapter plugs; each plug has an adapter type. A plug declaration is required to allow a function block to connect with a socket of the same adapter type in another function block.

E.1.5 Adapter socket list specification

SYNTAX:

```
'SOCKETS'                    <Sockets>
  (*socket declarations*)
'END_SOCKETS'                </Sockets>
```

This defines a list of adapter sockets; each socket has an adapter type. A socket declaration is required to allow a function block to accept a connection with a plug of the same adapter type in another function block.

E.1.6 Service declaration

SYNTAX:

```
'SERVICE' LeftInterfaceName '/' RightInterfaceName ':'
  (*sequence specifications*)
'END_SERVICE'

<Service LeftInterface="LeftInterfaceName"
    RightInterface="RightInterfaceName">
  (*sequence specifications*)
</Service>
```

This defines the dynamic interface behaviour provided by function block. The service has two parts to its name to reflect the initiating and receptive aspects of the service, e.g. `FBInterface/FBInternals`. The service body contains a list of service sequences that perform the various operations supported by the function block.

E.1.7 Service sequence declaration

SYNTAX:

```
'SEQUENCE' SequenceName
  (*service transactions*)
'END_SEQUENCE'

<ServiceSequence Name=SequenceName">
  (*service transactions*)
</ServiceSequence>
```

This defines a particular sequence of transactions to provide a given operation in a service specification. Each transaction is associated with an event arriving or issued across the resource boundary.

For example,

```
SEQUENCE event_delay
  E_DELAY.START(DT) -> E_DELAY.EO();
END_SEQUENCE
SEQUENCE delay_canceled
  E_DELAY.START(DT);
  E_DELAY.STOP();
END_SEQUENCE
```

```
<ServiceSequence Name="event_delay">
  <ServiceTransaction>
    <InputPrimitive Event="START" Interface="E_DELAY"
      Parameters="DT"/>
    <OutputPrimitive Event="EO" Interface="E_DELAY"/>
  </ServiceTransaction>
  <ServiceTransaction>
    <InputPrimitive Event="START" Interface="E_DELAY"
      Parameters="DT"/>
    <InputPrimitive Event="STOP" Interface="E_DELAY"/>
  </ServiceTransaction>
</ServiceSequence>
```

E.2 Basic function block type specification elements

For a basic function block, the function block type specification body contains, in addition to the interface specification, the definitions of internal variables, the execution control chart declaration, consisting of execution control states and transitions, and the algorithm declarations.

E.2.1 Internal variable specification

SYNTAX:

```
'VAR'                           <InternalVars>
  (*internal variables*)
'END_VAR'                       </InternalVars>
```

This is used to define internal variables required by the algorithms within a basic function block body.

E.2.2 Execution control state list specification

SYNTAX:

```
'EC_STATES'              (*not used in the XML format*)
  (*execution state declarations*)
'END_STATES'
```

This defines a list of execution control states that are used in the execution control chart (ECC) for a basic function block specification. Each state declaration contains a list of algorithms to be executed when the state is active and associated output events to be triggered when the algorithm has executed.

For example,

```
EVENT_STATES
  START;
  INIT : INIT -> INITO;
  POST : POST -> POSTO;
END_STATES
```

In the XML format, the execution control states are defined at the beginning of the <ECC> element used to encapsulate the overall execution control chart.

For example,

```
<ECC>
  <ECState Name="START"/>
  <ECState Name="INIT">
    <ECAction Algorithm="INIT" Output="INITO"/>
  </ECState>
  <ECState Name="POST">
    <ECAction Algorithm="POST" Output="POSTO"/>
  </ECState>
  (*transition list*)
</ECC>
```

E.2.3 Execution control transition list specification

SYNTAX:

```
'EC_TRANSITIONS'          (*not used in the XML format*)
  (*execution transition declarations*)
'END_TRANSITIONS'
```

This defines a list of execution control transitions that are used in ECC for a basic function block specification. Each transition has a transition condition. This is a logical expression that is expressed in terms of zero or one input event, input and output variables and internal variables and defines the condition to cause a state change from one given state to another.

For example,

```
EC_TRANSITIONS
  START TO INIT := INIT;
  INIT TO POST := [PV > 100 AND RUN];
END_TRANSITIONS
```

In the XML format, the execution control transitions are also defined within the <ECC> element.

For example,

```
<ECC>
  (*execution control states*)
  <ECTransition Source="START" Destination="INIT"
     Condition="INIT"/>
  <ECTransition Source="INIT" Destination="POST"
     Condition="[PV> 100 AND RUN]"/>
</ECC>
```

E.2.4 Algorithm declaration

SYNTAX:

```
'ALGORITHM' AlgorithmName 'IN' Language> ':'
  (*algorithm body*)
'END_ALGORITHM'

<Algorithm Name="AlgorithmName">
   (*enclosing elements define language and body for
   algorithm *)
</Algorithm>
```

This defines an algorithm within the body of a basic function block specification. A basic function block may have either no algorithm, or one or more algorithms. Each algorithm can be defined in a given language, e.g. C, ST, Java.

For example,

```
ALGORITHM POST IN ST:
  OUT := SP-PV;
  STATUS := 1;
END_ALGORITHM

<Algorithm Name="POST">
  <ST Text= "OUT := SP-PV; STATUS := 1;"/>
</Algorithm>
```

E.3 Composite function block type specification elements

For composite function blocks, the function block type specification body contains, in addition to the interface specification, a list of component function block instances, and a list of event, data and adapter connections between the components.

E.3.1 Function block instance list specification

SYNTAX:

```
'FBS'                    (*not used in the XML format*)
  (*internal function block instances*)
'END_FBS'
```

This defines a list of component function block instances required within a composite function block. Each function block instance name is given, followed by its function block type name.

For example,

```
FBS
  FB_INT : INTEGRAL_REAL;
  FB_DEV : DERIVATIVE_REAL;
END_FBS
```

In the XML format, component function block instances are defined at the beginning of the <FBNetwork> element used to encapsulate not only the component function blocks but also the event and data connections.

For example,

```
<FBNetwork>
  <FB Name="FB_INT" Type="INTEGRAL_REAL"/>
  <FB Name="FB_DEV" Type="DERIVATIVE_REAL"/>
  (*Event and data connection lists*)
</FBNetwork>
```

E.3.2 Connection list specification

SYNTAX:

```
(*event connection list*)
(*data connection list*)
(*adapter connection list*)
```

A connection list within the body of a composite function block specification can optionally contain a list of event connections, data connections and adapter connections. Note that a composite block may contain instances of basic function blocks connected by adapter plugs and sockets.

A connection list within an application or a subapplication is similar.

Event connection list specification

SYNTAX:

```
'EVENT_CONNECTIONS'                <EventConnections>
  (*event connection declarations*)
'END_CONNECTIONS'                  </EventConnections>
```

This defines a list of event connections between component function block inputs and outputs, and external interface events for a composite function block. A similar event connection list can be used in an application or in a subapplication.

For example,

```
EVENT_CONNECTIONS
  INIT TO FB_INT.INIT;
  FB_INT.OUT TO FB_DEV.PV;
  FB_INT.INITO TO INITO;
END_CONNECTIONS
```

```
<EventConnections>
  <Connection Source="INIT"
    Destination="FB_INT.INIT"/>
  <Connection Source="FB_INT.OUT"
    Destination="FB_DEV.PV"/>
  <Connection Source="FB_INT.INITO"
    Destination="INITO"/>
</EventConnections>
```

Data connection list specification
SYNTAX:

```
'DATA_CONNECTIONS'          <DataConnections>
  (*event connection declarations*)
'END_CONNECTIONS'           <DataConnections>
```

This defines a list of data connections between component function block inputs and outputs and the external interface. Each data declaration defines the mapping of data between component function block instances within a composite function block. A similar data connection list can be used in an application or a subapplication.

For example,

```
DATA_CONNECTIONS
  PV TO FB_INT.XIN;
  FB_INT.OUT TO OUT;
END_CONNECTIONS

<DataConnections>
  <Connection Source="PV" Destination="FB_INT.XIN"/>
  <Connection Source="FB_INT.OUT" Destination="OUT"/>
</DataConnection>
```

Adapter connection list specification
SYNTAX:

```
'ADAPTER_CONNECTIONS'       <AdapterConnections>
  (*adapter connection declarations*)
'END_CONNECTIONS'           </AdapterConnections>
```

This defines a list of adapter connections between function block inputs and outputs. Each adapter declaration defines the mapping between a plug and socket of component function block instances within a composite function block.

E.4 Subapplication type specification

SYNTAX:

```
'SUBAPPLICATION' SubappTypeName
  (*subapplication specification body*)
'END_SUBAPPLICATION'
```

```
<SubAppType Name="SubappTypeName">
  (*subapplication specification body*)
</SubAppType>
```

This is used to define a subapplication type specification. The body of each specification must contain the subapplication interface variable list. This identifies the input and output events and data passed across the subapplication interface.

A subapplication is built from a network of other function block and subapplication instances. The subapplication specification body contains a list of function block instance declarations, a list of subapplication instance declarations and a connections list.

E.4.1 Subapplication instance list specification

SYNTAX:

```
'SUBAPPS'                  (*not used in the XML format*)
  (*subapplication instance declarations*)
'END_SUBAPPS'
```

This is used to define a list of subapplication instances. Each subapplication instance name is given followed by its subapplication type. This is used within a subapplication or application specification body.

For example,

```
SUBAPPS
  SUB_Temp1 : LOOPController;
  SUB_Press1 : PressController;
END_SUBAPPS
```

Similar to the component function block instance declaration in composite function blocks, the XML format defines the subapplication instances directly in the function block network of applications and subapplications denoted with the `<SubAppNetwork>` element.

For example,

```
<SubAppNetwork>
  <SubApp Name="SUB_Temp1" Type="LOOPController"/>
  <SubApp Name="SUB_Press1" Type="PressController"/>
  (*Event and data connection lists*)
</SubAppNetwork>
```

E.5 Application type specification

SYNTAX:

```
'APPLICATION' ApplicationName <Application Name="ApplicationName">
  (*application specification body*)
'END_APPLICATION'            </Application>
```

This is used to define a complete application. The application specification body contains a list of function block instances, list of subapplication instances, an event connections list and a data connections list.

E.6 System configuration

SYNTAX:

```
'SYSTEM' SystemName              <System Name="SystemName">
  (*system configuration body*)
'END_SYSTEM'                     </System>
```

The system configuration body contains a complete definition of an IEC 61499 system. This contains application configuration details, device configurations and parameters and function block mappings.

Note: The Textual Syntax and XML format for `SYSTEM`, `RESOURCE` and `DEVICE` is rather complex; the reader is advised to consult the IEC 61499-1, Annex B and IEC 61499-1, Annex A for full and formal descriptions.

Appendix F
Bibliography

1. Nagel, R.N., and Dove, R.: '21st Century Manufacturing Enterprise Strategy: An Industry-Led View', Iacocca Institute, Lehigh University Press, Bethlehem, PA, 1991
2. Aguiar, M.W., Murgatroyd, I.S., and Edwards, J.M.: 'Object-oriented resource models: their role in specifying components of integrated manufacturing systems', *Computer Integrated Manufacturing Systems*, vol. 9. no. 1, pp. 33–48, 1996
3. Orfali, R., Harley, D., and Edwards, J.: *The Essential Distributed Objects Survival Guide*, Wiley, USA, 1996
4. Object Management Group, Data Distribution Service for Real-time Systems, Version 1.2, OMG Available Specification, 07-01-01
5. Josuttis, N.M.: *SOA in Practice: The Art of Distributed System Design*, O'Reilly Media, Sebastopol, CA, 2007
6. Luckham, D.: *The Power of Events: An Introduction to Complex Event Processing in Distributed Enterprise Systems*, Addison-Wesley, Boston, MA, 2002
7. Chappell, D.A.: *Enterprise Service Bus*, O'Reilly Media, Sebastopol, CA, 2004
8. Iwanitz, F., and Lange, J.: *OPC: Fundamentals, Implementation, and Application*, Hüthig Verlag, Heidelberg, 2006
9. Mahnke, W., Leitner, H.-S., and Damm, M.: *OPC Unified Architecture*, Springer, Berlin Heidelberg, 2009
10. International Electrotechnical Commission (2012): 'Function blocks – Part 1: Architecture', 2nd edn, IEC 61499-1 Ed. 2, IEC Geneva (also British Standard BS EN61499-1: 2012)
11. International Electrotechnical Commission (2013): 'Programmable controllers – Part 3: Programming languages', 3rd edn, IEC 61131-3 Ed. 3, IEC Geneva (also British Standard BS EN61131-3: 2013)
12. Lewis, R,: 'Programming industrial control systems using IEC 1131-3', *IEE Control Engineering Series* vol. 50, 1995
13. International Electrotechnical commission (2000): 'Programmable controllers – Part 5 communications', IEC 61131-5, IEC Geneva
14. International Electrotechnical Commission (2012): 'Function blocks – Part 2: Software tool requirements', 2nd edn, IEC 61499-2 Ed. 2, IEC Geneva (also British Standard BS EN61499-2: 2012)

15. International Electrotechnical Commission (2013): 'Function blocks – Part 4: Rules for compliance profiles', 2nd edn, IEC 61499-4 Ed. 2, IEC Geneva (also British Standard BS EN61499-4: 2013)
16. Czarnecki, K., and Eisenecker, U.: *Generative Programming: Methods, Tools, and Applications*, Addison-Wesley Professional, New York, NY, 2000
17. Szyperski, C.: *Component Software: Beyond Object-Oriented Programming*, 2nd edn, Addision-Wesely, Boston, MA,2002
18. Brambilla, M., Cabot, J., and Wimmer, M.: *Model-Driven Software Engineering in Practice*, Morgan & Claypool Publishers, USA, 2012
19. Kruchten, P.: 'The 4+1 view model of architecture', *IEEE Software*, November 1995
20. Zoitl, A.: *Real-Time Execution for IEC 61499*, Research Triangle Park, NC, ISA, 2009
21. *PLCopen Technical Committee 6, XML Formats for IEC 61131-3, Version 2.01 – Official Release*, PLCopen, May 2009
22. Pilone, D., and Pitman, N.: *UML 2.0 in a Nutshell*, O'Reilly, Sebastopol, CA, 2009
23. Selic, B., Gullekson, G., and Ward, P.T.: *Real-Time Object-Oriented Modeling*. Wiley Professional Computing, New York, 1994
24. Christensen, J.H., Strasser, T., Valentini, A., Vyatkin V., Zoitl, A., Carpanzano, E., *et al.*: The IEC 61499 *Function Block Standard Launch and Takeoff*, ISA Automation Week 2012, Asset Performance
25. Wenger, M., Zoitl, A., Sünder, C., Steininger, H., 'Transformation of IEC 61131-3 to IEC 61499 based on a model driven development approach.' Industrial Informatics, 2009 INDIN 2009. 7th IEEE Intl. Conf. on, pp. 715–720, 23–26 June 2009

Index

'2 out of 3 voting' algorithm 50–2
4+1 view model 16–17
4DIAC-IDE 202

ACCESS variables 11
actuators 188
adapter interface 79, 81, 85, 88
  developing 88
  in function block type 89
  textual syntax to 81–2
adapters
  in basic function blocks 83
  in composite function blocks 83
  concept 38, 79–81
  defining 81–2
  example of usage 152
  in function blocks 83
  guidelines for 88
  interface usage example 85–7
  modelling behaviour of 82
  in service interface function blocks 84–5
  structured interfaces with 80
  types 81–2
advanced distributed systems 4
agile manufacturing 1
airport baggage handling systems 162
algorithm, for function block behaviour 24, 55
algorithm execution 53, 58
AnalogInput function block type 148–9, 152
analogue input function block
  example 146–51
  generalisation of 151–3

application
  fractal form of 69
  general 22
  model 28–9
  structuring with subapplications 69–71
  timing and performance 32
  type specification 216
arrays 173
  data types 173
asynchronous behaviour 7

basic function blocks, adapters in 83
basic function block type 22, 53–62
  algorithm declaration 213
  execution control state list specification 211–12
  execution control transition list specification 212
  internal behaviour 53–5
  internal variable specification 211
  specification elements 211
bidirectional communication model 176
BOOL 170
Boolean 170
Boolean Falling Edge Detector 124
Boolean Rising Edge Detector 113, 123–4
business systems 3–4

CASE (Computer Aided Software Engineering) tools 13
Character set 165
CLIENT/SERVER function blocks 104, 105, 106
  Ethernet mapping 176–7

coherent data 7
'cold start' event 112
comments 166
commissioning 24
communication service interface function block 11, 103–6
Complex Event Systems 4
compliance profile 13, 159–60, 175–80
compliance to IEC 61499 155, 156, 159–60
component-oriented software 15
  relationship to function blocks 15–16
component function block 62, 63
composite function block 27, 62–8
  adapters in 83
  connection list specification 214
  definition 62
  example (sawtooth) 64–5
  function block instance list specification 213
  rules 62–3
  specification elements 213
  versus subapplication 183–4
conditional event propagation 127
configurability provisions 14, 160, 177–80
configuration constants 131
control tasks, modelling 128–9
conveyor
  distributed system model 141
  example network 138–9
ConveyorInterface adapter 86, 87, 88
Conveyor test station
  application of IEC 61499 135–41
  distributed system model 141–3
CORBA 4, 185
cyclic event chain example 113
Cyclic Event Generator 118–19

D (data latch) Bistable 123
data and event connections 186
data connection, rules for 63
data flow, concept in IEC 61499 7
data inputs
  for service interface function blocks 93–4
  function block interface 47–8
data outputs
  for service interface function blocks 94
  function block interface 47–8
data types 166–70
  use of 170–1
  in function block interface 47–8
dates and times of day data types 168–9
DCOM 185
DDS 4
Delayed Event Propagator 117–18
DerivativeReal function block 195–6
derived data type 171–3
development view 18
device
  model 30
  with multiple resources 32
  simple 160–1
  simple programmable 161
  in system model 29
  user-programmable 161
DeviceNet 143
DEV_MGR function block 177–8
direct digital control 134
Distributed Control System (DCS) 1–2, 18, 144
distributed function block 142
distributed industrial process measurement and control systems 5, 13, 21, 156, 182
distributed instrumentation 145
distribution model 32
  mapping of applications 32–4
  platform specific configurations 34
DKE (Deutsche Kommission Elektrotechnik Elektronik Informationstechnik) 161
DM_KRNL 178
DriveCntl function block 140
DriveCntl function block type 137
dynamic interface behaviour 46, 48–50

E_MERGE 65
EN and ENO concept 9–10
engineering costs 181
engineering support task 157–8
Enterprise Service Bus 4
enumerated data types 172
E_PERMIT function block 127
E_SPLIT 62, 63
Ethernet 5, 176, 177
event and data, difference between 186
event connection, rules for 63
Event-driven Bistable 122–3
event-driven model 24–7, 186–8
Event-driven Up counter 125
event flow, concept in IEC 61499 7
event flow control 125–7
event function blocks 111
  for event flow control 125–9
  graphical shorthand notation 126
  overview 111–12
  types 112–25
event inputs
  for service interface function blocks 93
  function block interface 47–8
Event Merger 114, 115
event outputs
  for service interface function blocks 93
  function block interface 47–8
Event Rendezvous 112, 114, 116
Event Selector 116–17
Event Splitter 114, 115
Event Switch 117, 123
Event Train Generator 119–20
Event Train Table-driven Generator 120–1
event type 187
exchange formats 39–43, 207–17
Exec function block type 137, 139
execution
  of algorithms 58
  of basic function blocks 56–9
  of composite function blocks 65
  of function blocks 24–7
Execution Control Chart (ECC) 27, 53–5, 135
  concept 46
  execution 56–9
  features 55–6
  notation 114
  primary states 58
'Execution Control' 23
execution model, for function blocks 24–7
execution timing 26
external interface declaration 46–52
external variables 184

FAQ (frequently asked questions), 181–9
Feed function block type 137–8, 140
Fermentation vessel 28
Fieldbus 3
  applications 143–53
  standardisation work 5
  standards 156
file exchange format 41, 158
Filter function block 149
floating point (REAL) data types 167
functional design 6
functional distribution 6
function block 22, 45
  adapter types in 83
  atomic execution 33
  basic 45–6, 53–62
  basic type 27
  composite type 27, 45–6, 62–8
  execution model for 24–7
  event function blocks 111–12
  external interface 45
  external interface declaration 46–52
  forms 45–6
  general characteristics 22–4
  instance name 22
  instances of basic function blocks 59
  main features 22
  management 106–8
  model 22
  network fragments 30

service interface type 27, 91
definition 98
as software component 23
types 27
instances 45
Function Block Development Kit (FDBK) 202
function block diagram (FBD) 8, 136
function block model
see function block
function block type 22, 45
adapter plug list specification 83–5, 209
adapter socket list specification 83–5, 209–10
data input/output specification 47–8, 208–9
definition 45, 207
event input/output specification 47–8, 208
service declaration 48–50, 210
service sequence declaration 48–50, 210
specification elements 207

GenAnalogInput function block type 152–3
global variables 10–11, 184

hardware costs 181
Heat1 function block 134
HiAlm function block 149
Human–Machine Interface (HMI) 2, 4, 8, 92
Hypertext Markup Language (HTML) 13

identifiers 165–6
IEC 61131-3 8, 135
ACCESS variables 11
communication function blocks 11
limitations 8
PLC software standard 158
relation to IEC 61499 5
software model 10
IEC 61158 143–4
IEC 61499
alternative graphical representations 184
application distribution 32–4, 183
benefits 181
communication sub-system 188
compliance 155, 159–61
cost of 185
criticism in 78
current status 155–7
development of 12–14
distributed system model 141
event-driven model 24–7, 186
execution control chart (ECC) 53–6, 184
faults and exceptions 127–8, 184
function block standard 5
global/external variables 23, 184
IEC technical working group 12
impact on IEC 61131 182
as methodology 5
object orientation questions 185
PAS 12, 157
sampled-data systems 187–8
stand-alone standard 182
'IEC 61499 Compliance Profile for Feasibility Demonstrations' 175
configurability provisions 177–80
interoperability provisions 176–7
portability provisions 175
industrial application, of IEC 61499 131
conveyor test station example 135–43
fieldbus applications 143–53
large-scale 161–2
temperature control example 132–5
industrially hardened PC (IPC) 3
INIT event 72–3, 94
INITO event 94
initial values 59, 171
Input1 function block 72
instance behaviour 59
integer data type 166–7
IntegralReal function block 197–8
interoperability provisions 13, 160, 176–7
Inventory function block type 138, 141

IO_READER 92, 100, 101, 127, 133, 139, 140, 141
  example 133
  function block 127
IO_WRITER 92, 100, 101, 128, 134, 140, 141
  example 134
IP (Internet Protocol) 143, 176, 179
ISO Technical Report 8509 48

Java 61, 195
  algorithms expressed using 201–2
  to express algorithm 53

Ladder Diagram 10
libraries, of function blocks 45, 158
'library element' 158
life-cycle 38
linear function blocks 152
literals 131
loading, function block fragments 36
logical view 17

management application 36
management commands 178–9
management function block 106–8
  main states 108
  service functions 107
management model 34–5
  applications 35–7
  operational state model of 37–9
ManValue input 149
mapping
  applications 32–4
  Fieldbus function blocks 146
mode function block 149
modes, modelling of 184

nxtControl 162, 202, 205
nxtSTUDIO 202, 203, 205

Object Management Group (OMG) 4
object-oriented (OO) software 14, 27
  concepts 185
  methodology 14
objects
  characteristics 14
  relationship to function blocks 15
OPC Foundation 4–5
operational state model 37–8
Output1 function blocks 72–3, 77

partnered service interface function blocks 97–8
permissive event, propagator 116, 117
physical view 18
PID (proportional, Integral and Derivative) 133–4, 201
  algorithm 191, 193
  component function blocks 194–5
  function block 72–3, 191–205
  in instruments 2
  as re-usable algorithm 22
PID_SIMPLE function block 133, 134–5
Piping and Instrumentation Diagram (P&ID) 6
plastics extruder 28
PLC (Programmable Logic Controllers) 1, 11, 41
  control system 3
  languages 8, 12, 30, 53
  soft PLC concepts 3
  software standard 158
  systems 2
PLCopen 41, 158
plugs 81, 83, 84, 143, 153, 207
  see also adapters
portability provisions 14, 160, 175
process control 14, 157, 187
process interface 23, 30, 181
process view 17–18
production line 2, 5, 28, 29
Profibus 143
Programmable Automation Controller (PAC) 3
Publicly Available Specification (PAS) 12, 157, 159
PUBLISH, example of usage 103, 104, 105, 141, 143

Publisher 76, 92
PUBLISH/SUBSCRIBE function blocks 103, 109, 176

QualStation function block type 138, 140

RAMP function block 61–2, 65
real-world applications, of IEC 61499 161–2
REQUESTER 98–100
resource
  model 30
  related to device 30
RESPONDER 99, 100
Restart Event Generator 119
RollOff function block type 138, 141
RSP input 99, 104, 105, 143
Run event 73, 138

sampled-data 187
Sawtooth function block, example 66
Scanner function block type 134–5
scenarios 17, 18
scheduling function 25, 31
semantic integration 5
sensors 188
Separate Event Table-driven Generator 121–2
Sequential Function Chart (SFC) 53–4, 59, 188
SERVICE 52, 102
service interface 27, 31–2
  function blocks
    adapters in 84–5
    for bidirectional transactions 104–6
    communication service interface 103–6
    examples 92, 100
    interface specification 93–8
    management function blocks 106–8
    overview 91–2
    partnered service interface 97–8
    textual syntax 101–3
    type definition 98–100
    types 27
    for unidirectional transactions 103–4
  service sequence diagrams 94–6
  specifications 93–8
  standard inputs and outputs for 93
service-oriented architectures (SOA) 4–5
service primitives 49–50, 85, 96, 188
service sequence diagrams 27, 48, 49, 50, 58, 59, 82, 118
  for application and resource triggered interactions 99
  behaviour of adapters 82
  for bidirectional transaction 105
  dynamic interface behaviour 48–50
  for partnered service interface function blocks 97
  for service interface function blocks 94–6
  for unidirectional transaction 104
  'Voter' example 51
service transactions 49–50, 97
SetPoint input 73
SI: *see* service interface
smart buildings 162
smart devices 4, 12, 162
smart grid 161–2
socket 79, 81, 83, 154, 207, 209
  see also adapters
software components 5, 6, 7, 15, 18, 19, 23
software tools 8, 14, 39, 114, 125, 155, 157, 160, 175, 177, 180, 182, 184, 186
  for testing function block models 202–5
STandard for the Exchange of Product data (STEP) 13
state-machine controllers 188, 189
static interface behaviour 46, 47–8
string data type 169–70

structured data 172
structured interfaces using adapters 78–88
see also adapters
Structured Text (ST) 53, 61, 135, 182, 195
subapplication
compared to composite function block 71
construction 71
distributed example 74–7
example 72–3
execution 72
feature of IEC 61499 74
and function block types 70
guidelines for 77–8
instance list specification 216
service interface function blocks 72–3
with structuring applications 69–71
TempControl 73, 76
type specification 215
sub-ranges data types 172–3
SUBSCRIBE
example of usage 141
function blocks 103, 104, 127, 141, 143, 176
subscriber 76, 92
syntax: *see* textual syntax
system configuration 14, 34, 39, 75, 77, 158, 217
system design 16, 75, 134, 154, 157, 161
system implementation time 181
system life cycle 38, 182
system model 29, 43
device model 29
overall system structure 29
resource model 30–2

TCP (Transmission Control Protocol) 176
technical integration 5
TempControl subapplication 73, 76
temperature control 4, 29
application of IEC 61499 132–5
example network 132
system 133
textual syntax 47, 61, 81, 83, 97, 155, 184, 207
composite example (sawtooth) 66
function block example 51–2
outline description 39–41
of PID function block 193–4
SI function block example 101
time data type 167–8
timing of events 112
TransSkt adapter 149, 151, 153
Two Event Rendezvous 114, 116

UDP (User Datagram Protocol) 176
unidirectional communication model 176

'vendor independent' design 161
Voter function block 50–2, 59, 60

WITH 47, 57, 66, 71, 81, 187
Wrt_Motor function block 139

XML 13, 39, 41, 42, 47, 158, 163, 180, 202, 216, 217
XML-exchange format 41–3, 207–17